Introduction to
Chemical Nomenclature

Introduction to Chemical Nomenclature

Fifth Edition

R. S. Cahn, MA, Dr Phil nat, FRIC, CC
*Formerly Editor to the Chemical Society and member of
the IUPAC Committee of Nomenclature of Organic Chemistry*

O. C. Dermer, BSc, PhD, ScD
*Regents Service Professor Emeritus of Chemistry,
Oklahoma State University*

BUTTERWORTHS
LONDON - BOSTON
Sydney - Wellington - Durban - Toronto

The Butterworth Group

United Kingdom	**Butterworth & Co (Publishers) Ltd**
London	88 Kingsway, WC2 6AB
Australia	**Butterworths Pty Ltd**
Sydney	586 Pacific Highway, Chatswood, NSW 2067
	Also at Melbourne, Brisbane, Adelaide and Perth
South Africa	**Butterworth & Co (South Africa) (Pty) Ltd**
Durban	152—154 Gale Street
New Zealand	**Butterworths of New Zealand Ltd**
Wellington	T & W Young Building, 77—85 Customhouse Quay, 1, CPO Box 472
Canada	**Butterworth & Co (Canada) Ltd**
Toronto	2265 Midland Avenue, Scarborough, Ontario, M1P 4S1
USA	**Butterworth (Publishers) Inc**
Boston	10 Tower Office Park, Woburn, Mass. 01801

First published 1959
Second edition 1964
Third edition 1968
Fourth edition 1974
Fifth edition 1979

© Butterworth & Co (Publishers) Ltd, 1979

ISBN 0 408 10608 5

British Library Cataloguing in Publication Data

Cahn, Robert Sidney
    Introduction to chemical nomenclature.
    — 5th ed.
    1. Chemistry — Nomenclature
    I. Title II. Dermer, Otis C
    540'.1'4   QD7   79-40303
    ISBN 0—408—10608—5

Typeset by Scribe Design, Medway, Kent
Printed in England by McCorquodale (Newton) Ltd
Newton-le-Willows, Lancs.

Preface

Since the previous edition of this 'Introduction to Chemical Nomenclature', not only has the International Union of Pure and Applied Chemistry (IUPAC) continued its publication of varied recommendations on nomenclature, but also Chemical Abstracts Service (CAS) has set out great changes which it has introduced into its indexing procedures. Bringing this 'Introduction' up to date would have been impossible for me had I not secured the collaboration of Professor O.C. Dermer of Oklahoma State University, Stillwater, Oklahoma, U.S.A., whose long experience of research and its publication and as section editor of *Chemical Abstracts (CA)* enabled the text to be modernized, corrected, and expanded so as to show the divergences between the recommendations of IUPAC and the indexing practices of *CA*. My indebtedness and thanks to him are boundless.

The principal expansions have been in the treatment of stereo-chemistry, natural products, and organometallic compounds.

It is a feature of many IUPAC recommendations that a choice of names is permitted; insofar as *CA* selects for its indexes and its computer services any one of the IUPAC offerings, it may be expected that a large body of chemists, particularly the younger members, will follow the more systematic *CA* usages; but when *CA* adopts completely new methods a more permanent conflict due to divergent loyalties may result. Professor Dermer and I agreed that the present book should merely set out the main differences between *CA* and IUPAC practices, only rarely stating a preference; this accords with the intention of previous editions, retained here, to explain the reasons rather than to dictate; and it is hoped that this will help the individual chemist who must know *CA* nomenclature as used in its indexes and information services as well as the IUPAC nomenclature that is seen in a great mass of current and past publications.

Professor Dermer and I are extremely grateful to S.P. Klesney (Secretary of the IUPAC Commission of Nomenclature of Organic Chemistry and in charge of the Central Report Index of Dow Chemical, U.S.A.) and Dr. W.H. Powell (of CAS and member of the above-mentioned IUPAC Commission), each of whom read the manuscript and made very many valuable corrections and additions. We are grateful also to IUPAC, Pergamon Press, and CAS for permission to quote from their publications, although unfortunately the IUPAC revised 1979 edition of Sections A, B, C, D, E, F and H of the organic nomenclature rules were not available when our manuscript went to press. Finally, we thank Mrs. Verne Allen Ospovat who with great efficiency typed the whole manuscript.

R.S. Cahn

Contents

1

The Development of Chemical Nomenclature

Nomenclature — the way that names are given to things — is one main point of difference between the language of chemistry, as of other sciences, and natural languages. The other is the importance of the written language compared with the spoken one. In chemistry there are several nomenclatures; not only elements and compounds must be named, but also reactions, methods, pieces of apparatus, and theoretical concepts. However, the vast numbers of compounds to be distinguished present the main problem, and the one addressed in the following pages.

Many things are designated by means other than names, and chemical compounds can be precisely represented by, for example, formulas, linear ciphers abbreviating formulas, or merely registry numbers. Such assemblies of symbols, not being words, are literally 'unspeakable' and are little discussed in this book. Even among words, chemists have some choice in referring to a compound; according to the occasion, '2-chloronaphthalene', 'that substance', or 'compound 2' may be the most suitable designation. Specific names, however, will continue to be needed for lists and legislation, as well as for abstracts, indexes, and lexicons. Of course names are essential, too, for most research papers or reports, for textbooks, and for chemical conversation whether written or oral.

The picturesque old appellations based on sources or properties of substances, or the name of the discoverer — spirits of hartshorn, muriatic acid, liver of sulfur, Glauber's salt, *etc.* — have long since disappeared from the chemist's language; and, as in any evolving language, the abandonment of old terms goes on. The need for reform was emphasized by Bergmann (ca. 1760), and it was met by the system of Guyton de Morveau, Lavoisier, and others (1787), based on the then novel idea that a name should indicate *composition*. This proved so very useful that before long it was widely accepted; but of course virtually all the well-characterized compounds of the time were inorganic, and relatively simple. Each such compound was later named by Berzelius as made up of an electropositive and an electronegative part, and two-word names of the kind he popularized are still familiar in inorganic nomenclature. (Berzelius also devised letter chemical symbols for the elements in much the form we use them today.) The success of the Berzelius names and philosophy, however, impeded the development of the concept of substitution in organic compounds (which it could not describe), and consequently that of substitutive nomenclature.

The rapid growth of organic chemistry in the latter part of the nine-teenth century produced a need for new systematization, especially for classification and indexing of compounds. This was undertaken by the Geneva Conference (1892). The rules developed there were seriously incomplete, but sound in principle, and are still used in the current edition of *Beilsteins Handbuch der organischen Chemie*. Since that time there have grown up two parallel efforts to improve and standardize chemical nomenclature, one by the International Union of Pure and Applied Chemistry (IUPAC) and the other by *Chemical Abstracts (CA)*. IUPAC Commissions for nomenclature of inorganic, organic, and biological chemistry were organized in 1922, and in other fields later; over the years, the first two have been most active in issuing recommen-dations. The need of *CA* to improve its indexes has led it to develop and publish rules, modified from time to time, by which it assigns index names, especially to chemical compounds.

It is the aim of systematic chemical nomenclature to describe the composition, and insofar as practicable the structure, of compounds. To the extent that this is achieved, chemists are fortunate; biologists, geologists, and astronomers have no such convenient way of associating scientific names with the things, or classes of thing, that they describe. However, no system of nomenclature can start afresh, abandoning all previous names. As a result, present practices are a patchwork, as diverse, specialized, and involved as the compounds they describe. Over the great complexities and illogicalities of current nomenclature hangs the shadow of the computer; the marshalling of four million structures and their attendant properties is increasingly admitted to be a computer responsibility. Because computers depend on logic, their use promotes systematic nomenclature. The groups of symbols best suited to computer programs to represent compounds, however, are mostly not names. Thus it does not appear that a computer language will soon displace the often arbitrary and sometimes inconsistent current usage. The origins of many common chemical names have been compiled in dictionary form[1].

A major shift to more structure-descriptive names has been made in recent subject indexes of *Chemical Abstracts* (*CA*), but there have been objections to replacing familiar names (such as *p*-benzoquinone) with longer ones (such as 2,5-cyclohexadiene-1,4-dione) that are not used in journals and books. Thus both the colloquial language, full of alternatives and irregularities, and the new one, with its complexities, will for many years, at least, still have to be read and understood. Nomenclature is thus worth study; every chemist should know its principles and its correct use.

The nomenclature that is today regarded as 'correct' is defined by the consensus of users' opinions. As in all linguistics, there is a struggle between the pragmatists, who regard as satisfactory any word that conveys the intended meaning, and the purists, who insist that rules ought to be followed, with the pragmatists having the advantage. Thus the Commissions of the International Union of Pure and Applied Chemistry (IUPAC) and of the International Union of Biochemistry (IUB) try to see nomenclature as a whole, codifying existing usage into rules and occasionally suggesting novelties; they accept the useful practices of specialists within their own fields but reject what they consider to be unnecessary aberrations from general principles. Since chemists differ widely in native language, and (as has been noted) the written language is more important to the chemist than the spoken, little attention is given to standardization of pronunciation.

While there are a very few problems in oral communication because two names have virtually the same sound (e.g., fluorine, fluorene), there are few of consequence that arise because two substances have the same written chemical name. It would be correspondingly simpler if there were only one 'correct' name for a substance, as in botany and zoology there is only one internationally authorized Latin name for a species of plant or animal. In chemistry, particularly organic chemistry, this is not so. The only exception is in indexes to collected works, where it is essential to place all entries under one name to save the user's time. Aside from the variations caused by differing natural languages, there are two reasons for the diversity. First, large compilations such as Beilstein's *Handbuch* and *CA* often use differing principles, and for them to introduce fundamental changes would bring chaos into their indexes; equally, a single rule is not always practicable, as when two or more large sections of chemists steadfastly maintain different customs. It is, however, noteworthy that *CA*, on the whole, conforms to the rules of the International Commissions, and conversely that those who formulate rules pay much attention to the practices of *CA*. There is one other feature that must be emphasized in a book such as this, devoted mainly to rules, namely that rules are a tool and not a master. Like other tools they can be used in different ways, or even set aside, or, better, modified when the science or its exposition is thereby improved or made easier to understand. This echoes Lavoisier's advice[2] of nearly 200 years ago: 'If languages are really instruments fashioned by men to make their thinking easier, they ought to be of the best kind possible; and to try to perfect them is actually to work for the advancement of science'.

Because, as just explained, it is not always possible for chemists to agree on the most desirable type of name, there are cases where alternative names are prescribed as equally 'correct' in the international rules. Then one country, Society, journal, or compendium may exercise its own preference. Within reason, each individual chemist has the same choice, though in practice he may be limited by his Society, editor, or publisher, and he is expected to be consistent in his choices. In most cases there is one name that is correct for a particular purpose: an author may use one of the alternatives, or even an unauthorized name, if it is essential for his theoretical arguments, but not just because of his personal preference; the authorized version will, with a little ingenuity, suffice for almost all purposes.

Now a systematic name for a complex compound is usually itself complex, and some thought will be needed to understand it. It is therefore misuse of nomenclature to scatter long chemical names indiscriminately into a cursive explanation of ideas. It is better to choose carefully a phrase such as 'the unsaturated alcohol', 'the derived acid', 'the starting material', or simply 'compound 5' (which has already been described by structure or name) than to bespatter one's prose with names such as 3-hydroxy-5-oxo-D-nor-5,6-secocholest-9(11)-en-6-oic acid or 5-(4-diethylamino-1-methylbutyl)dibenz[aj]acridine.

A more common misuse, which has produced some names now solidly entrenched, is false analogy in naming new types of compound, e.g., silicones ($R_2SiO)_x$ and sulfones (R_2SO_2) are very little like ketones (R_2CO) either structurally or chemically, but usage has made these names familiar. It is very hard to lay down precise rules for avoiding inappropriate names of this kind. Selection depends on a wide knowledge of previous practice: it is only too easy to mislead – and the overriding criterion for a name is that it shall be unambiguous. The advice of the national expert or editor is here essential.

Difficulties notwithstanding, chemists should, if they wish to be clearly understood, learn to describe accurately the compounds they are writing or talking about – and a definite act of learning is needed. Nomenclature, particularly in its modern developments, is not merely an arbitrary collection of names. It combines past practice with general principles, which it is the object of the following pages to explain. Tampering with it merely makes life harder for the reader and for the searcher in indexes. It is rarely good to call a spade a shovel, with or without a prefix.

REFERENCES

1. FLOOD, W.E., *The Origins of Chemical Names*, Oldbourne, London (1963)
2. LAVOISIER, A.L., *quoted by* SAVORY, T.H., *The Language of Science*, 2nd Ed., Deutsch, London (1966), p. 67

2

Inorganic

General

The greater part of inorganic nomenclature was for many years handled with reasonable ease by means of the endings '-ic', '-ous', '-ium', '-ide', '-ite', and '-ate'. When these did not suffice, help was sought mainly in prefixes of the type 'pyro-', 'hypo-', 'meta-', 'ortho-', 'per-', and 'sub-', and in endings such as '-oxylic', '-yl', and '-osyl'. There was, however, little consistency in the use of these adjuncts, and the resulting confusion was made worse when later studies of structure disclosed irrationalities in place of some of the supposed analogies. The Stock notation helped in many cases, and Werner's nomenclature was invaluable for coordination compounds.

There have been four international attempts in recent decades to devise a general system for inorganic nomenclature. A comprehensive set of rules was issued[1] by the Commission on the Nomenclature of Inorganic Chemistry of the International Union of Chemistry* in 1940, but because of the war it received no outside comment before publication. A revision was published in 1953 as 'Tentative Rules'[2]; independent comment and further consideration led to 'Definitive'† Rules[3] resulting from the Paris Conference of 1957. Some revisions were published in 1965[4]. Finally, in 1971 a new set of Definitive Rules[5] was published by IUPAC that amalgamates, revises, and greatly extends previous versions, providing principles, rules, and examples over a very wide range. It is on this last set, which has recently been summarized[6], that the present chapter is based. Among the chief features are acceptance of the well-known '-ide' nomenclature for binary compounds, recommendations for use of either the Stock or the Ewens–Bassett

*The older title, International Union of Pure and Applied Chemistry (IUPAC), was re-assumed in 1949.

†'Definitive' here means 'as accepted, or revised after being laid open to criticism by chemists'. It does not mean 'final, unalterable'.

notation, and extension of Werner's system for coordination compounds to a large part of general inorganic chemistry. Exceptions are still made for very common names such as water or ammonia and for a long list of acids, though the Commission doubtless hopes that these exceptions also will in time be superseded. Much that is familiar remains; and the extensions often lead to easily recognizable names such as potassium tetrachloroaurate(III) K [AuCl$_4$], hydrogen difluorodihydroxoborate H [B(OH)$_2$F$_2$], and potassium tetrafluorooxochromate(V) K [CrOF$_4$]; the extensions would, systematically, give disodium tetraoxosulfate for Na$_2$SO$_4$ though, of course, sodium sulfate is included among the permitted exceptions. The main virtue of the extension and revision is the replacement of personal or national preference by system and the provisions of unambiguous principles for naming new compounds, including many organometallic compounds of great complexity. It is unfortunate, however, that so many alternatives are left available.

Elements

Names and symbols for the elements are given in *Table 2.1* (pp. 8–9). The names in parentheses are those to be used with affixes, *e.g.*, cupric, ferrate. A few specific points may be noted. Tungsten is now accepted, after an earlier attempt to replace it by wolfram. The symbol for argon is Ar (not A), usage in different countries having become confused and the other noble gases having two-letter symbols. Some compounds of sulfur and antimony are named by use of syllables from the Greek (*thion*) or Latin (*stibium*); occasionally old French usage persists in English, as in azide from the French usage of *azote* for nitrogen. Use of wolframate and niccolate in place of tungstate and nickelate has been recommended [but *Chemical Abstracts* (*CA*) uses the latter older names]. Sulfur, not sulphur, should be used; the English use of sulphur is based on a mistaken belief that sulfur had a Greek origin, in which case ph would replace the Greek phi (ϕ). The American spellings cesium and aluminum may also be noted.

Naming of elements of atomic number greater than 105 on the basis of such numbers is recommended in tentative rules recently proposed by IUPAC[7]. This produces names such as unnilhexium (un-nil-hex-ium) for No. 106 and ununtrium for No. 113. It remains to be seen whether this unfamiliar system will replace naming new elements by scientists according to their personal preferences.

Some collective names now receive international sanction: noble gases; halogens (F, Cl, Br, I, At); chalcogens (O, S, Se, Te, Po); alkali metals (Li to Fr); alkaline-earth metals (Ca to Ra); lanthanoids for elements 57–71 (La to Lu inclusive) (lanthanides before 1965);

Table 2.1 IUPAC NAMES AND SYMBOLS OF THE ELEMENTS

Name	Symbol	Atomic number	Name	Symbol	Atomic number
Actinium	Ac	89	Lead (Plumbum)	Pb	82
Aluminum*	Al	13	Lithium	Li	3
Americium	Am	95	Lutetium	Lu	71
Antimony	Sb	51	Magnesium	Mg	12
Argon	Ar	18	Manganese	Mn	25
Arsenic	As	33	Mendelevium	Md	101
Astatine	At	85	Mercury	Hg	80
Barium	Ba	56	Molybdenum	Mo	42
Berkelium	Bk	97	Neodymium	Nd	60
Beryllium	Be	4	Neon	Ne	10
Bismuth	Bi	83	Neptunium	Np	93
Boron	B	5	Nickel	Ni	28
Bromine	Br	35	Niobium	Nb	41
Cadmium	Cd	48	Nitrogen	N	7
Calcium	Ca	20	Nobelium	No	102
Californium	Cf	98	Osmium	Os	76
Carbon	C	6	Oxygen	O	8
Cerium	Ce	58	Palladium	Pd	46
Cesium†	Cs	55	Phosphorus	P	15
Chlorine	Cl	17	Platinum	Pt	78
Chromium	Cr	24	Plutonium	Pu	94
Cobalt	Co	27	Polonium	Po	84
Copper (Cuprum)	Cu	29	Potassium	K	19
Curium	Cm	96	Praseodymium	Pr	59
Dysprosium	Dy	66	Promethium	Pm	61
Einsteinium	Es	99	Protactinium	Pa	91
Erbium	Er	68	Radium	Ra	88
Europium	Eu	63	Radon	Rn	86
Fermium	Fm	100	Rhenium	Re	75
Fluorine	F	9	Rhodium	Rh	45
Francium	Fr	87	Rubidium	Rb	37
Gadolinium	Gd	64	Ruthenium	Ru	44
Gallium	Ga	31	Samarium	Sm	62
Germanium	Ge	32	Scandium	Sc	21
Gold (Aurum)	Au	79	Selenium	Se	34
Hafnium	Hf	72	Silicon	Si	14
Helium	He	2	Silver (Argentum)	Ag	47
Holmium	Ho	67	Sodium	Na	11
Hydrogen	H	1	Strontium	Sr	38
Indium	In	49	Sulfur	S	16
Iodine	I	53	Tantalum	Ta	73
Iridium	Ir	77	Technetium	Tc	43
Iron (Ferrum)	Fe	26	Tellurium	Te	52
Krypton	Kr	36	Terbium	Tb	65
Lanthanum	La	57	Thallium	Tl	81
Lawrencium	Lr‡	103	Thorium	Th	90

Table 2.1 continued

Name	Symbol	Atomic number	Name	Symbol	Atomic number
Thulium	Tm	69	Vanadium	V	23
Tin (Stannum)	Sn	50	Xenon	Xe	54
Titanium	Ti	22	Ytterbium	Yb	70
Tungsten (Wolfram)	W	74	Yttrium	Y	39
			Zinc	Zn	30
Uranium	U	92	Zirconium	Zr	40

*Aluminium is nevertheless still current in British publications and is in accord
with the '-ium' ending adopted for all newly discovered elements.
†The spelling caesium or cæsium is generally used by British authors.
‡Not Lw as sometimes seen.

actinoids, uranoids, and curoids analogously. A transition element is
defined as an element whose atoms have an incomplete d subshell or
which gives rise to a cation or cations with an incomplete d subshell.
The term metalloid is vetoed: it is stated that elements should be
classified as metallic, semimetallic, or nonmetallic.

Protium, deuterium, and tritium are retained as names for the
hydrogen isotopes ^{1}H, ^{2}H, and ^{3}H, respectively, but other isotopes
should be distinguished by citing mass numbers, e.g., oxygen-18 or ^{18}O.
The prefixes are 'deuterio-' and 'tritio-' (not deutero-).

Indexes to be used with atomic symbols are:

left upper	. .	mass number
left lower	. .	atomic number
right upper	. .	ionic charge
right lower	. .	number of atoms

For example, $^{32}_{16}S_2^{2+}$ is a doubly charged molecule containing two atoms
of sulfur, each atom having the atomic number 16 and mass number 32.
The atomic number is obviously redundant and often omitted. Others
of these indexes may also be unnecessary; for instance, Ca^{2+} is a doubly
charged calcium ion (with natural abundance of isotopes), ^{15}N an
uncharged atom of nitrogen-15, $^{40}K^+$ a singly charged ion of potassium-
40. Although physicists formerly wrote the mass number as upper right
index, and many still do, the newer preference is for the positions given
above.

Radioactivity is often indicated by an asterisk, *K; it is rarely
necessary to give both the mass number and the asterisk (^{*40}K).

Ionic charge must be given as, e.g., superscript 2+, and not superscript +2.

For allotropic forms of elements a very simple numerical system is recommended: monohydrogen, dioxygen, tetraphosphorus, etc. Trioxygen is then recommended by IUPAC for O_3, though this can hardly be held to exclude use of the familiar name ozone since ozonide is listed among the recognized names of polyatomic anions.

Table 2.2 IUPAC DESCRIPTION OF SUBGROUPS

1A	2A	3A	4A	5A	6A	7A
K	Ca	Sc	Ti	V	Cr	Mn
Rb	Sr	Y	Zr	Nb	Mo	Tc
Cs	Pa	La*	Hf	Ta	W	Re
Fr	Ra	Ac†				

1B	2B	3B	4B	5B	6B	7B
Cu	Zn	Ga	Ge	As	Se	Br
Ag	Cd	In	Sn	Sb	Te	I
Au	Hg	Tl	Pb	Bi	Po	At

*Including the lanthanoids.
†Including the actinoids, but thorium, protactinium, and uranium may also be placed in groups 4, 5, and 6.

Ring and chain structures can be designated by prefixes *cyclo-* and *catena-*, e.g., *cyclo*-octasulfur (or octasulfur; for λ-sulfur), *catena*-sulfur (or polysulfur; for μ-sulfur). The prefix *cyclo-* is now specified for italics by IUPAC in inorganic chemistry (not in organic chemistry).

The description of subgroups of the Periodic Table has been settled by the IUPAC 1965 revision as shown in *Table 2.2.*

Compounds

Formulas and names should correspond to the stoichiometric proportions, expressed in the simplest form that avoids the use of fractions [though semi (½) and sesqui (1½) may be used for solvates and other addition compounds]. The molecular formula, if different, is used only when dealing with discrete molecules whose degree of association is considered independent of temperature. When there is temperature-dependence, the simplest formula is again to be used unless the molecular complexity requires particular emphasis in the context. Thus we have KCl potassium chloride, PCl_3 phosphorus trichloride, S_2Cl_2 disulfur dichloride, and $H_4P_2O_6$ hypophosphoric acid (*see Table 2.7*); NO_2 nitrogen dioxide represents the equilibrium mixture of NO_2 and N_2O_4 for normal use, but N_2O_4 dinitrogen tetraoxide is used where this doubling of the formula is significant.

In *formulas* the electropositive constituent is generally placed first, e.g., PCl_3, HCl. But there are exceptions, some merely by usage, e.g., NH_3 and NCl_3. When there is a central atom, that should normally be placed first with the remainder in alphabetical order, as in $PBrCl_2$ and PCl_3O; but $POCl_3$ is also authorized because PO may be considered a radical (*see* p. 23).

Names of compounds are given in two (or more) words, the (most) electropositive constituent (cation) first and the (most) electronegative (anion) last. Exceptions are made for neutral coordination compounds, addenda such as solvent molecules (*see* p. 38), and some hydrides. However, no fundamental distinction is to be made between ionized and non-ionized molecules in general.

Proportions of the various parts are expressed by Greek numerical prefixes (*see Table 3.1*, p. 46; also p. 67); but there are extremely important qualifications that mono (for unity) is usually omitted and that other numerical prefixes may also be omitted if no ambiguity results. Multiplicative numerical prefixes (bis, tris, tetrakis, etc.) are used when followed directly by another numerical prefix and may be used whenever ambiguity might otherwise be caused; and prefixes may be delimited by parentheses to aid clarity further (examples are on pp. 28–29). The terminal 'a' of tetra, penta, etc., was formerly elided in English before another vowel in inorganic chemistry, but this is expressly forbidden in IUPAC inorganic nomenclature: e.g., diphosphorus pentaoxide (not pentoxide).

Binary Compounds

Compounds between two elements are called binary compounds, independently of the number of atoms of each element in a molecule; e.g., they include N_2O, NO, NO_2, and N_2O_4.

In formulas and names of compounds between two nonmetals that constituent is placed first which occurs earlier in the sequence:

Rn, Xe, Kr, B, Si, C, Sb, As, P, N, H, Te, Se, S, At, I, Br, Cl, O, F

This order is arbitrary in places: it is not based solely on an order of electronegativity. The results are mostly familiar: NH_3 (not H_3N), CCl_4, NO, etc. But Cl_2O (chlorine monoxide) contrasts with O_2F (dioxygen fluoride).

When neither atom of a binary compound occurs in the sequence Rn . . . F above, the atoms are cited in the inverse order of the element sequence shown in *Table 2.3* (p. 12). This applies to both formulas and names, e.g., Na_2Pb, disodium plumbide.

Table 2.3 ELEMENT SEQUENCE

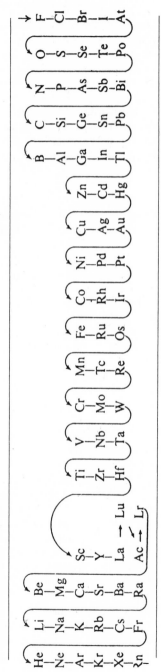

Metals precede nonmetals, in formulas and names.

When forming a complete name one leaves the name of the electro-positive constituent unmodified, except when it is necessary to indicate the valency or oxidation state (see below); and the same applies to the name of that one of a pair of nonmetals which is cited first in the list Rn . . . F. The name of the electronegative constituent, or of the non-metal named second, is modified to end in '-ide'. Normally this modification is carried out by stripping the name of the element back to the penultimate consonant and then adding '-ide', as in carbon, carb, carbide; or chlorine, chlor, chloride; but the following commonest exceptions should be noted:

bismuth, bismuthide	hydrogen, hydride
mercury, mercuride	nitrogen, nitride
oxygen, oxide	phosphorus, phosphide
zinc, zincide	

The group names chalcogenide and halogenide are also exceptions, but the term halide is more often seen.

When Latin names are given in *Table 2.1*, these are used, in the normal way.

Coupling this rule with those for numerals (*see* p. 11), and not forgetting the provisions for omission of numerical prefixes, we then have a host of familiar names of which the following list, based on one in the IUPAC rules, is representative: sodium chloride, calcium sulfide, lithium nitride, arsenic selenide, calcium phosphides, nickel arsenides, aluminum borides, iron carbides, boron hydrides, phosphorus hydrides, hydrogen chloride, silicon carbide, carbon disulfide, sulfur hexafluoride, chlorine dioxide, oxygen difluoride, sulfur dioxide, sulfur trioxide, carbon monoxide, carbon dioxide, dinitrogen oxide (N_2O), nitrogen oxide (NO), dinitrogen pentaoxide (N_2O_5).

This list will be seen to cover both ionized and un-ionized com-pounds. (Note the hyphen used to distinguish un-ionized from union-ized.) The compounds HHal are called hydrogen chloride, hydrogen bromide, hydrogen iodide, and hydrogen fluoride: the names hydro-chloric acid, etc., refer to aqueous solutions, and percentages such as 20 per cent hydrochloric acid denote the wt./wt. of hydrogen halide in the solution (not the extent of dilution of the concentrated acid by water). Also, hydrogen azide is recommended by IUPAC rather than hydrazoic acid.

Now, oxides could be, but are not, named by the coordination principle (*see* p. 17): e.g., SO_3 is sulfur trioxide (not trioxosulfur). The same applies to peroxides, i.e., compounds whose electronegative constituent is the group O_2^{2-} [when the electronegative constituents are atoms or ions of oxygen (O or O^{2-}) the compounds are dioxides] ; for example, Na_2O_2 sodium peroxide (but PbO_2 lead dioxide).

Binary hydrides may be named regularly, e.g., sodium hydride, tin tetrahydride. It is more usual, however, to name volatile hydrides by adding -ane to the root name of the non-hydrogen element, e.g., borane (BH_3), stannane (SnH_4). If the number of atoms of the non-hydrogen element in the molecule exceeds one, this is indicated by a Greek numerical prefix, as in diborane (B_2H_6), trisilane (Si_3H_8), and pentasulfane (H_2S_5). There is a short list of exceptions hallowed by long usage, namely: water, ammonia, hydrazine, phosphine, arsine, stibine, and bismuthine. Also hydrides of Group VII, being acidic, are named as hydrogen halides (cf. above). Occasionally hydride names ending in -ane may become confused with the organic termination -ane for a six-membered ring; for this reason, IUPAC recommends disilane H_2Se_2 and ditellane H_2Te_2 and their homologues.

Many trivial names have been abandoned in modern nomenclature, more often owing to mere omission from IUPAC rules than to prohibition there. Notable cases are the oxides of nitrogen.

For many compounds, in many circumstances, it is unnecessary to specify valence or oxidation state. It can always be dispensed with when no ambiguity results, as can the numerical prefixes. So one still has very many simple names such as sodium chloride, calcium oxide, deuterium oxide, and barium chloride.

However, there are far more cases where such simplicity leaves ambiguous names and at this stage the concept of oxidation number must be stressed. It is expressed in the 1970 IUPAC rules as follows:

'The oxidation number of an element in any chemical entity is the charge which would be present on an atom of the element if the electrons in each bond to that atom were assigned to the more electronegative atom (as exemplified in *Table 2.4*).

'By convention hydrogen is considered positive in combination with nonmetals. The conventions concerning the oxidation numbers of organic radicals and the nitrosyl group are considered below.

'In the elementary state the atoms have oxidation state zero and a bond between atoms of the same element makes no contribution to the oxidation number (as in *Table 2.5*).'

When given elements can combine in different proportions the names describing the resulting different compounds may be formed in four ways:

(1)   The first way is to use numerical prefixes that show the stoichiometric composition, as in iron trichloride, copper dichloride, triiron tetraoxide, and dinitrogen pentasulfide.

Table 2.4 EXAMPLES OF OXIDATION NUMBERS

				Oxidation numbers	
MnO_4^-	=	one Mn^{7+} and four O^{2-} ions	Mn = VII	O = −II	
ClO^-	=	one Cl^+ and one O^{2-} ion	Cl = I	O = −II	
CH_4	=	one C^{4-} and four H^+ ions	C = −IV	H = I	
CCl_4	=	one C^{4+} and four Cl^- ions	C = IV	Cl = −I	
NH_4^+	=	one N^{3-} and four H^+ ions	N = −III	H = I	
NF_4^+	=	one N^{5+} and four F^- ions	N = V	F = −I	
AlH_4^-	=	one Al^{3+} and four H^- ions	Al = III	H = −I	
$[PtCl_2(NH_3)_2]$	=	one Pt^{2+} and two Cl^- ions and two uncharged NH_3 molecules	Pt = II	Cl = −I	
$[Ni(CO)_4]$	=	one uncharged Ni atom and four uncharged CO molecules	Ni = 0		

Table 2.5 FURTHER EXAMPLES OF OXIDATION NUMBERS

				Oxidation numbers	
P_4	=	four uncharged P atoms	P = 0		
P_2H_4	=	two P^{2-} and four H^+ ions	P = −II	H = I	
C_2H_2	=	two C^- and two H^+ ions	C = −I	H = I	
O_2F_2	=	two O^+ and two F^- ions	O = I	F = −I	
$Mn_2(CO)_{10}$	=	two uncharged Mn atoms and ten uncharged CO molecules'	Mn = 0		

(2)   In the second method the oxidation number, in Roman numerals in parentheses, is placed immediately after the name of the element concerned, virtually always the electropositive one (but zero is denoted by the Arabic 0). This is the Stock system*, which is widely favoured. Examples are:

MnO_2 manganese(IV) oxide
PbO   lead(II) oxide

The rules list both English and Latin names and favour the latter where both are given in *Table 2.1*:

$FeCl_3$ iron(III) chloride or ferrum(III) chloride
$CuCl_2$ copper(II) chloride or cuprum(II) chloride

but this preference is not followed in English–American writing.

*In IUPAC rules and many other places, Stock numbers are written in small capitals in parentheses, e.g., lead(IV). In some cases, however, large capitals are used. It is highly desirable that the oxidation number should be enclosed in parentheses, and that these should be close up to the word they qualify [e.g., lead(IV) *not* lead (IV) or lead IV]

(3)   In ionic compounds the charge on an ion may be indicated by an Arabic numeral, followed by the sign of the charge, both in parentheses attached at the end of the name of the ion (the Ewens–Bassett system, now used in *CA* indexes). Examples are:

$FeCl_3$ iron(3+) chloride

$CuCl_2$ copper(2+) chloride

(4)   There is of course also the much older system of denoting a higher valence state by the ending -ic, and a lower one by the ending -ous. This system is notoriously incomplete when an element can have more than two valence states, is not applicable to coordination compounds, and can be a troublesome tax on memory for the less common elements. The IUPAC Commission lists the system as 'in use but discouraged', although there is also a statement that it 'may be retained for elements exhibiting not more than two valencies'.

Here we may note that it is only in connection with the '-ous'/'-ic' and '-ite'/'-ate' terminations, and with the definition of coordination entity. that the word 'valency' is used in the 1970 IUPAC rules; elsewhere their rules are based on oxidation numbers and ionic charges.

Pseudobinary Compounds

This name has been applied to compounds containing some common polyatomic anions that customarily have names ending in '-ide'; these anions are shown in *Table 2.6*. Treating these like the other '-ide' names

Table 2.6   NAMES FOR PSEUDOBINARY ANIONS

HO^-	hydroxide ion	N_3^-	azide ion
O_2^{2-}	peroxide ion	NH^{2-}	imide ion
O_2^-	hyperoxide* ion	NH_2^-	amide ion
O_3^-	ozonide ion	$NHOH^-$	hydroxylamide ion
S_2^{2-}	disulfide ion	$N_2H_3^-$	hydrazide ion
I_3^-	tri-iodide ion	CN^-	cyanide ion
HF_2^-	hydrogen difluoride ion	C_2^{2-}	acetylide ion

Chemical Abstracts uses superoxide.

we obtain NaOH sodium hydroxide, KCN potassium cyanide, and the like. Sodium amide ($NaNH_2$) is sometimes abbreviated to sodamide, but this irregular practice is undesirable.

The list above includes S_2^{2-} as the disulfide ion and there are clearly homologous ions; but in this series it has become the practice to name the chain compounds $HS-[S]_x-SH$ as binary hydrides, i.e., sulfanes

(x = 1, trisulfane; x = 2, tetrasulfane; etc.). This is particularly valuable for organic derivatives and for hydroxy, amino, etc., derivatives, but it is then probably wisest to name the substituents only as prefixes rather than to encounter the difficulties in deciding a choice of suffix.

The Extended Coordination Principle

In its oldest sense, the sense in which it is usually understood, a coordination compound is one containing an atom (A) directly attached to other atoms (B) or groups (C), or both, the number of these being such that the oxidation number or stoichiometric valence of (A) is exceeded.

The 1957 IUPAC rules extended this principle by abolishing the restriction that the oxidation number must be exceeded. The effect is to bring the major part of nomenclature potentially within the scope of the nomenclature customary for coordination compounds. For instance, in the group SO_4, the sulfur atom is A and the oxygen atoms are B. In diatomic groups the choice of the central atom may pose a little difficulty, but oxygen rarely serves as such; thus in ClO^- the Cl is central atom.

For discussion of this nomenclature some definitions should be noted (some will not be needed until later in this chapter, but it is convenient to group them together):

Central or nuclear atom: the atom A above.

Coordinating atom: each atom *directly* attached to the central atom.

Coordination number or ligancy: the number of atoms directly attached to a central atom.

Ligand: each atom (B) or group (C) attached to the central atom.

Multidentate ligand: a group containing more than one *potential* coordinating atom. Hence, unidentate, bidentate, etc.

Chelate ligand: a ligand actually attached to one central atom through two or more coordinating atoms.

Bridging group: a ligand attached to more than one central atom.

Complex: the whole assembly of one or more central atoms with their attached ligands.

Polynuclear complex: a complex containing more than one central (nuclear) atom. Hence, mononuclear, dinuclear, etc.

To name a complex, the names of the ligands are attached directly in front of the name of the central atom; the oxidation number of the central atom is then stated last, or the proportions of the constituents are indicated by stoichiometric numerical prefixes; when the oxidation number is exceeded, both the numerical prefixes and the oxidation number are often needed.

A complex may be cationic, neutral, or anionic. Names of cationic and neutral complexes are not modified (except for a few mentioned below), i.e., they have no characteristic ending. Names of anionic complexes are given the ending '-ate', often with abbreviation.

It is very necessary to realize that according to this system the ending -ate denotes *merely* an anionic complex. When an element has variable oxidation state this use of -ate does *not* denote a higher oxidation state; it is used with all oxidation states, the distinction between them being made by citing the oxidation number. Particular note must be taken of this because the '-ate', '-ite', '-ide' distinction is still permitted in a limited and defined, but large, number of cases, and the two nomenclatures are likely to be in use side by side for some time.

The ligands are to be cited in alphabetical order.* In this alphabetization, multiplying prefixes are neglected; e.g., di-iodo is alphabetized under i and thus after fluoro, but both are placed after trichloro, which is alphabetized under c (for further details *see* p. 63).

The names of anions as ligands are changed to end in -o (*see* p. 27); those of neutral and cationic ligands are unchanged, except that H_2O is denoted aqua† and NH_3 ammine.

Acids and Normal Salts Containing More Than Two Elements

Classical coordination compounds, that is, those where the oxidation number of the central atom is exceeded, are deferred to p. 26.

(1) The practices mentioned earlier hold, about citing electropositive before electronegative constituents, using Stock numbers or '-ic', '-ous' endings, and omitting numerical prefixes, or use of the Ewens–Bassett system, or Stock numbers, when no confusion arises.

For acids containing more than two elements the IUPAC recommendations of 1970 are somewhat more permissive than those of 1957. The latter held out hope that the nomenclature typified by sulfuric acid would be replaced by hydrogen sulfate and the like. The 1970 rules freely permit both types, while, however, expressing preference for the 'hydrogen -ate' type for the less common acids.

Both sets of rules list common acids for which the 'acid' nomenclature is likely to be retained. *Table 2.7* is the more extensive 1970 list.

---

*The 1957 rules gave complex instructions about the order of citation of ligands. All that is now swept away in favour of the alphabetical order.
† Aqua is a change from aquo, which was customary before 1965.

Table 2.7 NAMES FOR ACIDS CONTAINING MORE THAN TWO ELEMENTS

H_3BO_3	orthoboric acid† or boric acid	H_2SO_4	sulfuric acid
$(HBO_2)_n$	metaboric acid	$H_2S_2O_7$	disulfuric acid
H_2CO_3	carbonic acid	H_2SO_5	peroxomonosulfuric acid
HOCN	cyanic acid		
HNCO	isocyanic acid	$H_2S_2O_8$	peroxodisulfuric acid
HONC	fulminic acid	$H_2S_2O_3$	thiosulfuric acid*
H_4SiO_4	orthosilicic acid†	$H_2S_2O_6$	dithionic acid
$(H_2SiO_3)_n$	metasilicic acid	H_2SO_3	sulfurous acid
HNO_3	nitric acid	$H_2S_2O_5$	disulfurous acid
HNO_4	peroxonitric acid*	$H_2S_2O_2$	thiosulfurous acid*
HNO_2	nitrous acid	$H_2S_2O_4$	dithionous acid
HOONO	peroxonitrous acid	H_2SO_2	sulfoxylic acid
H_2NO_2	nitroxylic acid	$H_2S_xO_6$	polythionic acids
$H_2N_2O_2$	hyponitrous acid	($x = 3,4...$)	
H_3PO_4	orthophosphoric† or phosphoric acid	H_2SeO_4	selenic acid
		H_2SeO_3	selenious acid
$H_4P_2O_7$	diphosphoric or pyrophosphoric acid	H_6TeO_6	orthotelluric acid†
		H_2CrO_4	chromic acid
		$H_2Cr_2O_7$	dichromic acid
$(HPO_3)_n$	metaphosphoric acid	$HClO_4$	perchloric acid
		$HClO_3$	chloric acid
H_3PO_5	peroxomonophosphoric acid	$HClO_2$	chlorous acid
		HClO	hypochlorous acid
$H_4P_2O_8$	peroxodiphosphoric acid	$HBrO_4$	perbromic acid
		$HBrO_3$	bromic acid
$(HO)_2OP-PO(OH)_2$	hypophosphoric acid or diphosphoric(IV) acid	$HBrO_2$	bromous acid
		HBrO	hypobromous acid
$(HO)_2P-O-PO(OH)_2$	diphosphoric(III,V) acid	H_5IO_6	orthoperiodic acid†
		HIO_4	periodic acid
H_2PHO_3	phosphonic acid	HIO_3	iodic acid
$H_2P_2H_2O_5$	diphosphonic acid	HIO	hypoiodous acid
HPH_2O_2	phosphinic acid	$HMnO_4$	permanganic acid
H_3AsO_4	arsenic acid	H_2MnO_4	manganic acid
H_3AsO_3	arsenious acid	$HTcO_4$	pertechnetic acid
$HSb(OH)_6$	hexahydroxo-antimonic acid	H_2TcO_4	technetic acid
		$HReO_4$	perrhenic acid
		H_2ReO_4	rhenic acid

*Certain other peroxoacids and thioacids, as well as selenoacids and telluroacids, can be named similarly.

†Ortho- need be used only when distinction from other acids is essential.

Anions from the acids in *Table 2.7* are formed by changing '-ous acid' to '-ite', and '-ic acid' to '-ate', and this gives the familiar names for a further host of common compounds. For instance, $NaNO_2$ sodium nitrite and $NaNO_3$ sodium nitrate follow at once. Na_2SO_4 sodium sulfate and Na_2CO_3 sodium carbonate follow similarly when permission to omit unnecessary prefixes (e.g., from disodium sulfate) is remembered.

A few general principles for the use of prefixes embodied in *Table 2.7* may be pointed out. A prefix hypo- denotes a lower oxidation state and is used with the '-ous' and the '-ic' terminations. 'Per-' denotes a higher oxidation state, is used only with the '-ic' acids (and their salts), and must not be confused with 'peroxo-' (*see* below). 'Ortho-' and 'meta-' distinguish acids of differing water content.

Thioacids are acids derived by replacing oxygen in the derived anion by sulfur, and are named by adding 'thio-' before the trivial name of the acid; and selenoacids and telluroacids are treated similarly. Examples are: $H_2S_2O_2$ thiosulfurous acid; $H_2S_2O_3$ thiosulfuric acid; KSCN potassium thiocyanate; H_3PO_3S monothiophosphoric acid; $H_3PO_2S_2$ dithiophosphoric acid; H_2CS_3 trithiocarbonic acid. However, $H_2S_2O_6$ dithionic acid and $H_2S_2O_4$ dithionous acid (not hydrosulfurous or hyposulfurous) in *Table 2.7* are exceptional to this treatment, and higher homologues are named analogously.

Peroxoacids, in which $-O-$ is replaced by $-O-O-$, are similarly distinguished by the prefix 'peroxo-' ('peroxy-' or simply 'per-' have frequently been used in the past), as, for example, in HNO_4 peroxonitric acid and K_2SO_5 potassium peroxosulfate, $HO-O-NO$ peroxonitrous acid, and $H_4P_2O_8$ peroxodiphosphoric acid.

(2)   To explain how acids and salts are named by the coordination method requires, first, the principles given in the preceding section and then further precision about endings.

It will be remembered that the electropositive portion is named before the electronegative, as usual, and that the electronegative, anionic portion has an ending '-ate'. This ending is normally added to the penultimate consonant of the name of the central atom (the Latin name being used if one is given in *Table 2.1*), as with the '-ide' ending; but the exceptions include the following: antimony, antimonate; bismuth, bismuthate; carbon, carbonate; cobalt, cobaltate; nickel, nickelate*; nitrogen, nitrate; phosphorus, phosphate; wolfram, wolframate; zinc, zincate.

Names for ligands (which in the simple acids now considered are all anionic) are formed by changing the endings '-ide', '-ite', or '-ate' to '-ido', '-ito', or '-ato', respectively, for direct union to the name of the central atom. The names shown in *Table 2.8* do not conform but are retained in deference to common older usage.

In this way $H_2[SO_4]$ becomes dihydrogen tetraoxosulfate(VI); or, since the 'acid' terminology is also permitted, this might become tetraoxosulfuric acid. If we were to make the further provision that oxo-prefixes might be omitted when no confusion is caused, we should reach the familiar sulfuric acid; nevertheless, such argument can lead to confusion in some cases, so it is better to regard the name sulfuric acid simply as one of the permitted exceptions.

*IUPAC rules recommend niccolate.

Table 2.8 NAMES RETAINED FOR ANIONIC LIGANDS

F^-	fluoro	O^{2-}	oxo
Cl^-	chloro	H^-	hydrido or hydro*
Br^-	bromo	OH^-	hydroxo
I^-	iodo	O_2^{2-}	peroxo†
HO_2^-	hydrogenperoxo	CN^-	cyano
S^{2-}	thio	CH_3O^-	methoxo† or
(but: S_2^{2-} disulfido)			methanolato
HS^-	mercapto	CH_3S^-	methylthio or
			methanethiolato

*Both hydrido and hydro are used for coordinated hydrogen but the latter term is usually restricted to boron compounds.
†In conformity with the practice or organic nomenclature, the forms peroxy and methoxy are also used but are not recommended

The majority of simple oxoacids are included in *Table 2.7*. Note the names orthoperiodic acid for H_5IO_6 and periodic acid for HIO_4, which resolve previous differences of opinion.

Some examples of real usefulness of the method are:

$HReO_4$	tetraoxorhenic(VII) acid
H_3ReO_5	pentaoxorhenic(VII) acid
H_2ReO_4	tetraoxorhenic(VI) acid
$HReO_3$	trioxorhenic(V) acid
$H_4Re_2O_7$	heptaoxodirhenic(V) acid
H_3GaO_3	trioxogallic(III) acid
H_4XeO_6	hexaoxoxenonic(VIII) acid

In simpler cases, it seems reasonable to omit the 'oxo-' prefixes, as in manganic(VI) acid for H_2MnO_4, and manganic(V) acid for H_3MnO_4.

So far as simple compounds are concerned, it is perhaps with halo-acids that the Stock nomenclature comes best into its own. Potassium hexachloroplatinate(IV) for K_2PtCl_6 sidesteps the older platinichloride. Chlorosulfuric(VI) acid for HSO_3Cl may be regarded as an abbreviated form of chlorotrioxosulfuric(VI) acid; it should *not* be named chloro-sulfonic acid by the substitutive method of organic chemistry. According to this system the hydrogen of acids is named as the cation, and it is this principle that leads to 'hydrogen -ate' names. Some further simple examples may be here adduced as illustrations:

$K[AuCl_4]$	potassium tetrachloroaurate(III)
$Na[PHO_2F]$	sodium fluorohydridodioxophosphate
$H[PF_6]$	hydrogen hexafluorophosphate
$H_4[XeO_6]$	hydrogen hexaoxoxenonate(VIII)
	(cf. above)

H [B(OH)$_2$F$_2$]	hydrogen difluorodihydroxoborate
Na$_4$ [Fe(CN)$_6$]	sodium hexacyanoferrate(II)
K [AgF$_4$]	potassium tetrafluoroargentate(III)
Ba [BrF$_4$]$_2$	barium tetrafluorobromate(III)
K [Au(OH)$_4$]	potassium tetrahydroxoaurate(III)
Na [AlCl$_4$]	sodium tetrachloroaluminate
Li [AlH$_4$]	lithium tetrahydridoaluminate

The last example in that list may come as a shock to many organic chemists, for it replaces the older and less precise lithium aluminum hydride.

Some further points in the above examples may also repay comment. The most recent IUPAC rules recommend that in formulas the square brackets indicating the extent of a complex ion be separated by a space from any other chemical symbol or another set of square brackets. Current British and American books and journals ignore this recommendation, which is perhaps regarded as painting the lily — after all, the square brackets themselves define the extent of the complex ion.

The prefix 'hydrido-' denoting a hydride anionic ligand (*see* p. 21) is especially noteworthy; it follows the general rule of changing an ending '-ide' to '-ido' and for inorganic chemistry is considered preferable to the organic usage of 'hydro'. A similar argument could have been applied to the other exceptional anionic ligand prefixes listed on p. 21, e.g., chlorido- in place of chloro-, but here the need to alter previous usage was not so necessary as with hydrogen, which so often occurs as a proton H$^+$

Lastly, it may be well to anticipate here a little from later pages by mentioning that hydrocarbon radicals that are present as ligands do not receive the terminal '-o'; so we have, for example, Na [B(C$_6$H$_5$)$_4$] sodium tetraphenylborate.

Ions and Radicals

Anions are negatively, and cations positively, charged atoms or groups of atoms. Radicals are uncharged groups of atoms that occur (not necessarily always filling the same role) throughout a number of compounds and do not normally exist in the free state; when they exist independently and uncharged they are 'free radicals'; when charged they become ions. Names of ions are, in general, those of the electropositive or electronegative 'constituents' described in the preceding sections; names of cations thus have no characteristic ending; those of anions end in '-ide', '-ite', or '-ate'. Thus Na$^+$ is the sodium ion or, more precisely, the sodium cation. Fe^{2+} is the ferrous ion, or iron(II) ion, or ferrum(II) ion (or 'cation' may replace 'ion'). I^- is the iodide ion or iodide anion, so it obviously is not wise to abbreviate the name of the less common I^+ iodine cation.

There are a few special points to be noted, particularly where the IUPAC rules choose between previous alternative usages.

NO^+ is to be called the nitrosyl cation, NO_2^+ the nitryl cation (not nitroxyl, to avoid confusion with the radical from nitroxylic acid).

Polyatomic cations formed by adding more protons to monoatomic anions than are required to give a neutral unit have the ending '-onium': ammonium, phosphonium, arsonium, oxonium (H_3O^+), sulfonium, selenonium, telluronium, iodonium. Substituted derivatives may be formed from them, e.g., methoxyammonium, tetramethylstibonium, dimethyloxonium $(CH_3)_2OH^+$. Current *CA* names for nitrogenous cations do not follow these rules: *see* p. 121).

For nitrogen bases other than ammonia and its substitution products the cations are named by changing the final '-e' to '-ium': anilinium, imidazolium, glycinium, etc.; $N_2H_5^+$ is hydrazinium(1+); $N_2H_6^{2+}$ is hydrazinium(2+). Uronium and thiouronium (from urea and thiourea) are exceptions. Names such as dioxanium and acetonium are formed analogously.

Finally, though H^+ is the proton and H_3O^+ (a monohydrated proton) is oxonium, the term hydrogen ion can be used when the degree of hydration is of no importance in the particular circumstances.

Table 2.9   SPECIAL INORGANIC RADICAL NAMES

HO	hydroxyl	SeO	seleninyl
CO	carbonyl	SeO_2	selenonyl
NO	nitrosyl	CrO_2	chromyl
NO_2	nitryl	UO_2	uranyl
PO	phosphoryl	NpO_2	neptunyl
SO	sulfinyl (thionyl)	PuO_2	plutonyl*
SO_2	sulfonyl (sulfuryl)	ClO	chlorosyl†
S_2O_5	disulfuryl	ClO_2	chloryl†
		ClO_3	perchloryl†

*Similarly for other actinides.
†Similarly for other halogens.

Radicals having special names are listed in *Table 2.9*. 'Thio-', 'seleno-', etc., prefixes may be used with these, as with acids. The '-yl' principle is noted in the rules as not extensible to other metal–oxygen radicals. In some cases use of Stock numbers or the Ewens–Bassett system extends the range of utility of these special radical names, e.g., UO_2^{2+} uranyl(VI) or uranyl(2+), UO_2^+ uranyl(V) or uranyl(1+).

These radical names can be used to construct compound names such as:

$COCl_2$	carbonyl chloride
PON	phosphoryl nitride
$PSCl_3$	thiophosphoryl chloride
CrO_2Cl_2	chromyl chloride
IO_2F	iodyl fluoride
SO_2NH	sulfonyl imide or sulfuryl imide

It will be seen that these '-yl' radical names are always handled as if they were the electropositive part of a name, but no polarity considerations should in fact apply: NOCl is nitrosyl chloride, and $NOClO_4$ is nitrosyl perchlorate, independently of views on polarity.

The radical names, then, are convenient for certain groups of compound, but must be restricted to compounds consisting of discrete molecules, and other, superficially similar, compounds are often better named as mixed oxides or oxide salts (*see* p. 26).

The IUPAC rules use these radicals also for amides, giving $SO_2(NH_2)_2$ sulfonyl diamide and $PO(NH_2)_3$ phosphoryl triamide. As alternatives they give sulfuric diamide and phosphoric triamide. For the partial amides they put forward '-amidic acid' names, and as an alternative the use of coordination nomenclature; for example, NH_2SO_3H sulfamidic acid or amidosulfuric acid. Abbreviated names — in the above cases the common sulfamide, phosphamide, and sulfamic acid — are 'not recommended'. However, the '-yl amide' names differ noticeably from present organic practice, where amides are a much larger class than in inorganic chemistry; it seems to the present authors better to use either the trivial names or the coordination names at present: for example, $SO_2(NH_2)_2$ sulfamide; or, for new compounds, as in diamidodioxo-sulfur. *Chemical Abstracts* uses sulfamide and sulfamic acid as index headings.

Salts and Salt-like Compounds

(1) 'ACID' SALTS, i.e., SALTS CONTAINING ACID HYDROGEN

'Names are formed by adding the word "hydrogen", with numerical prefix where necessary, to denote the replaceable hydrogen in the salt. Hydrogen shall be followed without space by the name of the anion. Exceptionally, inorganic anions may contain hydrogen which is not replaceable. It is still designated by hydrogen, if it is considered to have the oxidation number +1, but the salts cannot of course be called acid salts. Examples are:

(1)	$NaHCO_3$	sodium hydrogencarbonate
(2)	LiH_2PO_4	lithium dihydrogenphosphate
(3)	KHS	potassium hydrogensulfide
(4)	$NaHPHO_3$	sodium hydrogenphosphonate'

That is a direct transcription from the 1970 IUPAC rules. The citation of acidic hydrogen is classical nomenclature, but the junction of two words (hydrogen and an anion) that can have independent existence, without any form of punctuation, is rare in British or American nomenclature. Its justification is that the primary ions present are, e.g., Na^+ and $[HCO_3]^-$, but the custom has not yet become popular; it is not used by *CA*.

It is to be noted that names such as bicarbonate and bisulfate are not permitted: nothing similar is anyway possible for polybasic acids, the historical basis for these old names is unfamiliar, and the word hydrogen should always be cited when this atom is present. Derived ions should be similarly named, e.g., hydrogencarbonate ion or hydrogen carbonate ion. Again many English-writing chemists have been unwilling to conform by giving up these 'bi-' names.

Example (4) in the list above, sodium hydrogenphosphonate, emphasizes that the name phosphonic acid is given to H_2PHO_3 (*see Table 2.7, p. 19*) although this acid was still given the older name phosphorous acid in the 1957 IUPAC rules. Phosphonic acid is in line with organic nomenclature such as phenylphosphonic acid for $C_6H_5PO_3H_2$ (*see p. 170*), and the IUPAC rules for inorganic compounds specify the name phosphonate for the anion PHO_3^{2-}. However, the analogous sulfur ion HSO_3^-, which is known to exist mainly as the structure in which the H is attached to S, is still named as the oxygen tautomer $HOS(O)O^-$, hydrogensulfite (or hydrogen sulfite), rather than as sulfonate $HS(O)_2O^-$.

(2) DOUBLE SALTS, etc.

A somewhat complex set of IUPAC rules published in 1957 has since been much simplified by extensive use of the alphabetical order.

In formulating and in naming double salts, etc., all the cations are now to be cited first, in alphabetical order, except that H^+ is named last. Then all the anions are listed, these also in their alphabetical order. Water considered to be coordinately bound to a specific ion is cited as aqua-. Numerical prefixes can be omitted if the oxidation states are constant or defined. Examples are:

$NaTl(NO_3)_2$	sodium thallium(I) nitrate *or* sodium thallium dinitrate
$NaZn(UO_2)_3(C_2H_3O_2)_9 \cdot 6H_2O$	sodium triuranyl zinc acetate hexahydrate
$Na_6ClF(SO_4)_2$	(hexa)sodium chloride fluoride bis(sulfate)
$[Cr(H_2O)_6]\,Cl_3$	hexaaquachromium(III) chloride *or* hexaaquachromium trichloride

Note the use of bis(sulfate) to avoid confusion with disulfate $(S_2O_7^{2-})$.

(3) 'BASIC' SALTS

These are to be named as double salts with O^{2-} or OH^- anions; names such as hydroxychloride join the constituents together incorrectly and should not be used (in English). The IUPAC class names are oxide salts and hydroxide salts, e.g.,

$Mg(OH)Cl$	magnesium chloride hydroxide
$BiOCl$	bismuth chloride oxide
$Cu_2(OH)_3Cl$	dicopper chloride trihydroxide

(4) DOUBLE OXIDES AND HYDROXIDES

These are named on similar principles, the metals being cited in alphabetical order. Examples are:

$FeTiO_3$	iron(II) titanium(IV) trioxide
$Cu(CrO_2)_2$	chromium(III) copper (II) oxide
$AlLiMn_2O_4(OH)_4$	aluminum lithium dimanganese(IV) tetrahydroxide tetraoxide

(Copper chromite is considered incorrect for the second example in view of results of structural studies on the solid.)

Classical Coordination Compounds

These are the compounds where the oxidation number is exceeded. Most of the principles governing their nomenclature have been explained

above. The main features may be recapitulated as follows. The complex
may be cationic, neutral, or anionic. Names for cationic or neutral complexes have no special ending (except '-ium'; *see* p. 23); names of anionic
complexes end in -ate. Ligands are cited in alphabetical order. Stock
numbers or Ewens–Bassett charge numbers are cited when necessary or
helpful, but stoichiometric prefixes may in some cases suffice alone.

Cationic and neutral ligands have unaltered names; anionic ligands
are given the ending '-o'.

These and the following additional principles suffice for a very large
number of coordination compounds:

(1)   Hydrocarbon radicals have the usual organic names ending in '-yl'
(not '-ylo'), but they are treated as anions when computing the oxidation
number.
(2) CH_3O^-, etc., are called methoxo, etc.
(3) CH_3S^- is named methanethiolato or methylthio.
(4)   When an organic compound that is not normally named as an acid
forms a ligand *by loss of a proton* it is treated as anionic and its name
is given the ending '-ato'. (In other cases it is treated as neutral.) When
necessary the charge on such a ligand is to be stated, e.g.,

$-O_2CCH(O^-)CH(OH)CO_2^-$ tartrato(3−)
$-O_2CCH(OH)CH(OH)CO_2^-$ tartrato(2−)

(5)   For computation of the oxidation number, NO, NS, CO, and CS
are treated as neutral.
(6)   Water and ammonia as neutral ligands are designated aqua and
ammine, respectively.
(7)   When a ligand might be attached at different points, distinction is
made by means of element symbols -*S*, -*N*, -*O*, etc., at the end of the
name, unless the customary name makes this clear (e.g., −SCN
thiocyanato, −NCS isothiocyanato).
(8)   Names of neutral ligands, other than H_2O, NH_3, NO, NS, CO, and
CS, are placed in parentheses, and these are preceded by any necessary
multiplicative numerical prefixes (bis, tris, etc.)
(9)   Enclosing marks are used in the sequence { [( )] } for composite
groups, as necessary, but the order is [{ ( )}] if the outside square
brackets define the complex.

The examples in *Table 2.10* are selected from IUPAC lists[5] and will,
it is hoped, be self-explanatory and sufficiently illustrate these principles
of the alternative systems. Bold type illustrates operation of the alphabetical order (for details *see* pp. 18 and 63). Names of some of the organic
ligands have been changed since the IUPAC rules were compiled, but the

Table 2.10 EXAMPLES OF NAMES FOR CLASSICAL COORDINATION COMPOUNDS

Na [B(NO₃)₄]
sodium tetranitratoborate(1−)
sodium tetranitratoborate(III)

[Co(N₃)(NH₃)₅] SO₄
pentaammineazidocobalt(2+) sulfate
pentaammineazidocobalt(III) sulfate

[Ru(HSO₃)₂(NH₃)₄]
tetraamminebis(hydrogensulfito)ruthenium
tetraamminebis(hydrogensulfito)ruthenium(II)

K [B(C₆H₅)₄]
potassium tetraphenylborate(1−)
potassium tetraphenylborate(III)

[Fe(C₂C₆H₅)₂(CO)₄]
tetracarbonylbis(phenylethynyl)iron
tetracarbonylbis(phenylethynyl)iron(II)

[Ni(C₄H₇N₂O₂)₂]
bis(2,3-butanedione dioximato)nickel
bis(2,3-butanedione dioximato)nickel(II)

[CoCl₂(C₄H₈N₂O₂)₂]
bis(2,3-butanedione dioxime)dichlorocobalt
bis(2,3-butanedione dioxime)dichlorocobalt(II)

[CuCl₂(CH₃NH₂)₂]
dichlorobis(methylamine)copper
dichlorobis(methylamine)copper(II)

[Pt(py)₄][PtCl₄]
tetrakis(pyridine)platinum(2+) tetrachloroplatinate(2−)
tetrakis(pyridine)platinum(II) tetrachloroplatinate(II)

[Cr(H₂O)₆] Cl₃
hexaaquachromium(3+) chloride
hexaaquachromium trichloride

[Co(NH₃)₆] Cl(SO₄)
hexaamminecobalt(3+) chloride sulfate
hexaamminecobalt(III) chloride sulfate

[CoCl₃(NH₃)₂{(CH₃)₂NH}]
diamminetrichloro(dimethylamine)cobalt
diamminetrichloro(dimethylamine)cobalt(III)

continued

Table 2.10 (*continued*)

K [Co(CN)(CO)$_2$(NO)]
potassium dicarbonylcyanonitrosylcobaltate(1−)
potassium dicarbonylcyanonitrosylcobaltate(0)

[CoH(N$_2$){(C$_6$H$_5$)$_3$P}$_3$]
(dinitrogen)hydridotris(triphenylphosphine)cobalt
(dinitrogen)hydridotris(triphenylphosphine)cobalt(I)

[NiCl$_3$(H$_2$O){N(CH$_2$CH$_2$)$_3$$\overset{+}{\text{N}}CH_3$}]
aquatrichloro{1-methyl-4-aza-l-azoniabicyclo[2.2.2]octane}nickel
aquatrichloro{1-methyl-4-aza-1-azoniabicyclo[2.2.2]octane}nickel(II)

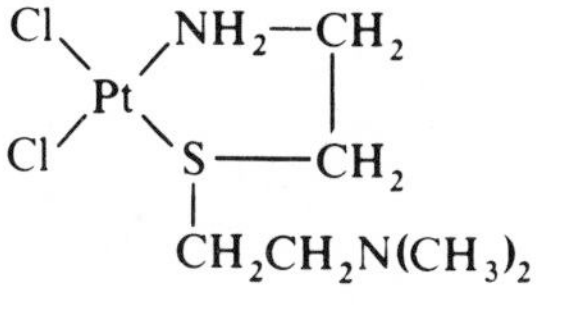

dichloro[*N,N*-dimethyl-2,2′-thiobis(ethylamine)-*S,N′*]platinum
dichloro[*N,N*-dimethyl-2,2′-thiobis(ethylamine)-*S,N′*]platinum(II)

---

general principles of this inorganic nomenclature are not affected
thereby.

The 1970 IUPAC rules give much further useful information on how
to name the extremely complicated structures that arise in this ever-
growing class of compound. It cannot be fully reproduced in this intro-
ductory text, but the examples cited in this chapter give an idea of its
range.

(1) ABBREVIATIONS

A valuable list of abbreviations recommended by IUPAC[5] for common
ligands is reproduced in *Table 2.11* (p. 30).

(2) π-COMPLEXES

In some circumstances it suffices to give only the stoichiometric composition of π-complexes, e.g.,

[PtCl$_2$(C$_2$H$_4$)(NH$_3$)]
amminedichloroethyleneplatinum

Table 2.11 IUPAC ABBREVIATIONS FOR SOME COMMON LIGANDS

Hacac	acetylacetone, 2,4-pentanedione, $CH_3COCH_2COCH_3$
acac	acetylacetonato
Hbg	biguanide, $H_2NC(=NH)NHC(=NH)NH_2$
H_2dmg	dimethylglyoxime, 2,3-butanedione dioxime, $CH_3C(=NOH)C(=NOH)CH_3$
Hdmg	dimethylglyoximato(1−)
dmg	dimethylglyoximato(2−)
H_4edta	ethylenediaminetetraacetic acid, $(HO_2CCH_2)_2NCH_2CH_2N(CH_2CO_2H)_2$
H_2ox	oxalic acid, HO_2C-CO_2H
bpy	2,2′-bipyridine or 2,2′-bipyridyl, 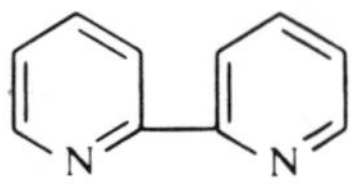
diars	o-phenylenebis(dimethylarsine), $(CH_3)_2AsC_6H_4As(CH_3)_2$
dien	diethylenetriamine, $H_2NCH_2CH_2NHCH_2CH_2NH_2$
diphos	ethylenebis(diphenylphosphine), $Ph_2PCH_2CH_2PPh_2$
en	ethylenediamine, $H_2NCH_2CH_2NH_2$
phen	1,10-phenanthroline 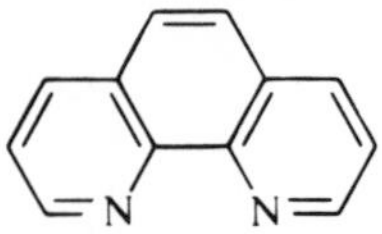
pn	propylenediamine, $H_2NCH(CH_3)CH_2NH_2$
py	pyridine
tren	2,2′,2″-triaminotriethylamine, $(H_2NCH_2CH_2)_3N$
trien	triethylenetetraamine, $(H_2NCH_2CH_2NHCH_2)_2$
ur	urea, $(H_2N)_2CO$

When some or all unsaturated atoms in a chain or ring are bound to the central atom, structure may be indicated by use of the Greek letter η, which may be read as eta or hapto (from the Greek *to fasten*). The compound $[PtCl_2(C_2H_4)(NH_3)]$ would then be named:

 amminedichloro(η-ethylene)platinum   *or*
amminedichloro(η-ethylene)platinum(II)

Locants are used as necessary; they are not needed when all skeletal atoms are bound to the central atom, but *CA* then usually adds a superscript or locant to show their number or position and omits the Ewens–Bassett number for neutral species, as, for example, in:

$Cr(C_6H_6)(CO)_3$
(η-benzene)tricarbonylchromium
(η-benzene)tricarbonylchromium(0)
(η^6-benzene)tricarbonylchromium *(CA)*

A further selection of examples may be instructive:

$[ReH(C_5H_5)_2]$
bis(η-cyclopentadienyl)hydridorhenium
bis(η-cyclopentadienyl)hydridorhenium(III)

$[Co(C_5H_5)(C_5H_6)]$
(η-cyclopentadiene)(η-cyclopentadienyl)cobalt
(η-cyclopentadiene)(η-cyclopentadienyl)cobalt(I)

tetracarbonyl(η-1,5-cyclooctadiene)molybdenum
tetracarbonyl(η-1,5-cyclooctadiene)molybdenum(0)

(1–3-η-2-butenyl)tricarbonylcobalt
(1–3-η-2-butenyl)tricarbonylcobalt(I)

tricarbonyl(1–4-η-cyclooctatetraene)iron

trans-μ-(1–4-η:5–8-η-cyclooctatetraene)bis(tricarbonyliron) (for μ see next section)

Fe(C$_5$H$_5$)$_2^-$
bis(η-cyclopentadienyl)iron
bis(η-cyclopentadienyl)iron(II)
ferrocene

[Fe(C$_5$H$_5$)$_2$] [BF$_4$]
bis(η-cyclopentadienyl)iron(1+) tetrafluoroborate
bis(η-cyclopentadienyl)iron(III) tetrafluoroborate
ferrocene(1+) tetrafluoroborate(1−)
ferrocenium tetrafluoroborate

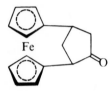

2,4-(1,1′-ferrocenediyl)cyclopentanone

Ferrocene may be used as a parent, but proliferation of '-ocene' names is not favoured by IUPAC.

(3) BRIDGING GROUPS

These are designated by the Greek letter μ (mu) prefix and are followed by a hyphen (this hyphen is desirable to differentiate, e.g., μ-dichloro- from nonbridging dichloro atoms occurring in the same complex:

[(NH$_3$)$_5$Cr−OH−Cr(NH$_3$)$_5$] Cl$_5$
μ-hydroxo-bis[pentaamminechromium)(5+) chloride
μ-hydroxo-bis[pentaamminechromium(III)] chloride

[(CO)$_3$Fe(CO)$_3$Fe(CO)$_3$]
tri-μ-carbonyl-bis(tricarbonyliron)

[Br$_2$Pt(SMe$_2$)$_2$PtBr$_2$]
bis(μ-dimethyl sulfide)-bis[dibromoplatinum(II)]

[(CO)$_2$Ni(Me$_2$PCH$_2$CH$_2$PMe$_2$)$_2$Ni(CO)$_2$]
bis[μ-ethylenebis(dimethylphosphine)] -bis(dicarbonylnickel)

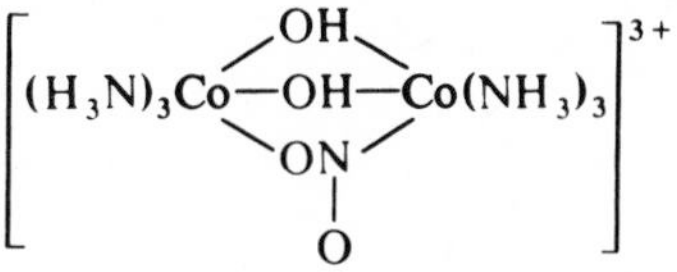

hexaammine-di-μ-hydroxo-μ-nitrito(O,N)-dicobalt(3+) ion
hexaammine-di-μ-hydroxo-μ-nitrito(O,N)-dicobalt(III) ion

$[Be_4 O(CH_3 CO_2)_6]$
hexa-μ-acetato-(O,O')-μ_4-oxo-tetraberyllium
hexa-μ-acetato-(O,O')-μ_4-oxo-tetraberyllium(II)

$[Cr_3 O(CH_3 CO_2)_6]$ Cl
hexa-μ-acetato-(O,O')-μ_3-oxo-trichromium(1+) chloride
hexa-μ-acetato-(O,O')-μ_3-oxo-trichromium(III) chloride

(4) EXTENDED STRUCTURES

Many compounds of extended (polymeric) structure exist; their nature
is identified by a prefix *catena-*, as in:

$[Cs]_n [\cdots CuCl_2 - Cl - CuCl_2 - Cl - CuCl_2 - Cl \cdots]^{n-}$

cesium *catena*-μ-chloro-dichlorocuprate(II)

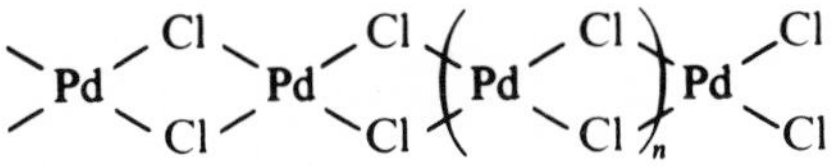

catena-di-μ-chloro-palladium

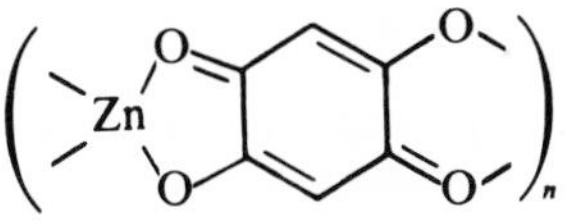

catena-μ-[2,5-dioxido-*p*-benzoquinone(2−)-*O,O'$:$O'',O'''*]-zinc
catena-μ-[2,5-dioxido-*p*-benzoquinone(2−)-*O,O'$:$O'',O'''*]-zinc(II)

(5) DI- AND POLY-NUCLEAR COMPOUNDS

When there is no bridging group, symmetrical di- and poly-nuclear
compounds are named by use of bis-, etc., but for unsymmetrical

compounds one component (the last in *Table 2.3*) is treated as substituted by the others, as in:

$[Br_4 Re-ReBr_4]^{2-}$
bis(tetrabromorhenate)(2−)
bis[tetrabromorhenate(III)]

$[(C_6H_5)_3 AsAuMn(CO)_5]$
pentacarbonyl[(triphenylarsine)aurio] manganese

When there are also bridging groups, or otherwise when necessary, the metal–metal bond is indicated by italicized symbols at the end of the name, e.g.:

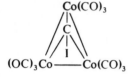

μ_3-iodomethylidyne-*cyclo*-tris(tricarbonylcobalt) (3 *Co–Co*)

The interfix *triangulo* may be used here instead of *cyclo*; *see* next section.

$Os_3(CO)_{12}$
cyclo-tris(tetracarbonylosmium) (3 *Os–Os*)

An alternative name for the last compound is supplied in the next section.

(6) HOMOATOMIC AGGREGATES (CLUSTERS)

The geometrical shape of the cluster is indicated by an abbreviated affix of geometrical significance, e.g. *triangulo*-, *quadro*-, *tetrahedro*-, *octahedro*-, *dodecahedro*-, etc. Examples are the following, but this type of geometrical labeling seems likely to be overwhelmed as such names proliferate:

$Os_3(CO)_{12}$
dodecacarbonyl-*triangulo*-triosmium

B_4Cl_4
tetrachloro-*tetrahedro*-tetraboron

$[Mo_6 Cl_8 Cl_6]^{2-}$
octa-μ_3-chloro-hexachloro-*octahedro*-hexamolybdate(2−) ion
octa-μ_3-chloro-hexachloro-*octahedro*-hexamolybdate(II) ion

(7) ISOMERISM

Prefixes *cis-* and *trans-* are used for 4-planar and 6-octahedral complexes, supplemented by *fac-* (facial) and *mer-* (meridional), as illustrated in *Figure 2.1*.

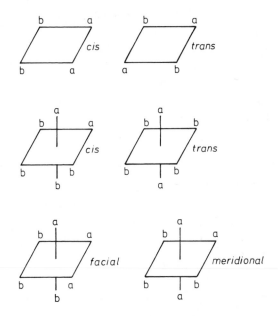

Figure 2.1. Prefixes used for 4-planar and 6-octahedral complexes

In many cases locants are needed to distinguish isomers; the choice of locants is based in the IUPAC rules on locating planes of atoms perpendicular to a major axis and is described in elaborate rules. For octahedral complexes the method described in the sequence-rule paper[8] is an alternative but does not have approval of the IUPAC Commission for the Nomenclature of Inorganic Chemistry. However, a notation based on the Cahn–Ingold–Prelog sequence rule is now used by *CA* to form part of the names for mononuclear coordination compounds[9].

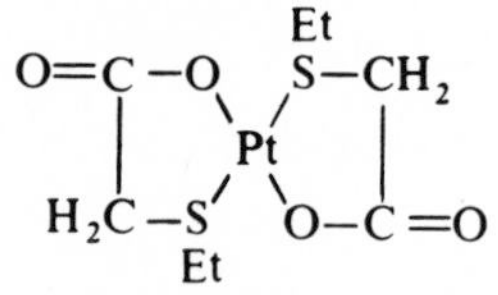

O=C—O   S—CH₂

$$O{=}C{-}O \quad S{-}CH_2$$

(structure diagram)

Pt

H₂C—S   O—C=O
Et

trans-bis[(ethylthio)acetato-*O,S*] platinum
trans-bis[(ethylthio)acetato-*O,S*] platinum(II)

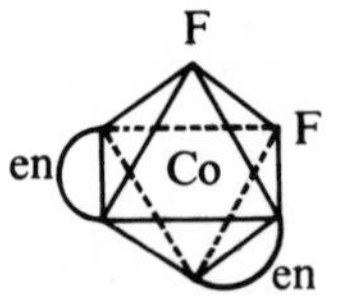

cis-bis(ethylenediamine)difluorocobalt(1+) ion
cis-bis(ethylenediamine)difluorocobalt(III) ion

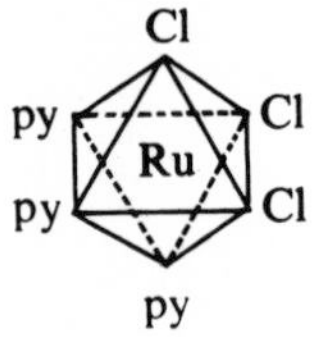

fac-trichlorotris(pyridine)ruthenium
fac-trichlorotris(pyridine)ruthenium(III)

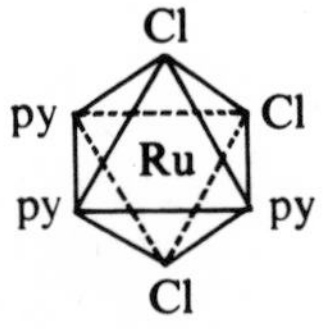

mer-trichlorotris(pyridine)ruthenium
mer-trichlorotris(pyridine)ruthenium(III)

(8) ABSOLUTE CONFIGURATION

The sequence rule[8] can be used for describing the absolute configuration of tetrahedral and octahedral complexes; it prescribes (*R*), (*S*) or (*P*), (*M*) symbols for identification. For six-coordinate complexes with tris- and bis-bidentate ligands, however, the IUPAC Inorganic Commission recommends a different method that leads to symbols Λ, Λ (or δ, λ for conformations).

Miscellaneous

ISOPOLYANIONS

Numerical prefixes, together with Stock numbers or Ewens–Bassett ion charges, usually suffice:

$K_2S_2O_7$	potassium disulfate
$Ca_3Mo_7O_{24}$	tricalcium heptamolybdate
$[O_2HP-O-PHO_2]^{2-}$	dihydrogendiphosphate(III)(2−) (trivial name diphosphonate)
$[O_2HP-O-PO_3H]^{2-}$	dihydrogendiphosphate(III,V)(2−)

Prefixes *cyclo-*, *catena-*, or μ- are used when it is felt to be desirable. Some examples are:

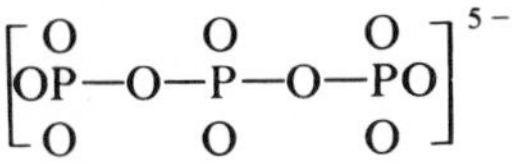

triphosphate

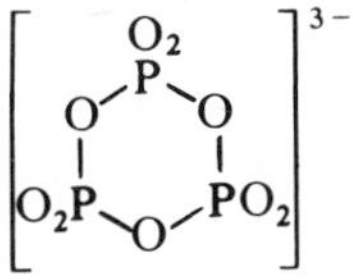

cyclo-triphosphate

$[O(PO_3)_n]^{(n+2)-}$ *catena*-polyphosphate

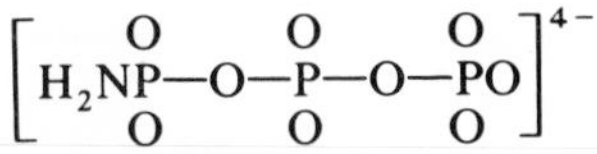

1-amidotriphosphate(4−)

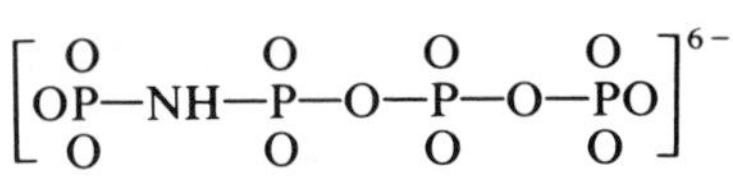

1,2-μ-imidotetraphosphate(6−)

HETEROPOLYANIONS

Linear heteropolyanions are named by treating the terminal component that comes last in alphabetical order as having the others as ligands. If the terminal units are the same, choice rests on the second from the end, and so on:

$[O_3P-O-SO_3]^{3-}$	phosphatosulfate(3−)
$[O_3Cr-O-AsO_2-O-PO_3]^{4-}$	(chromatoarsenato)phosphate(4−)

Cyclic heteropolyanions receive the prefix *cyclo-*; alphabetical order decides the starting point and direction of citation, as in:

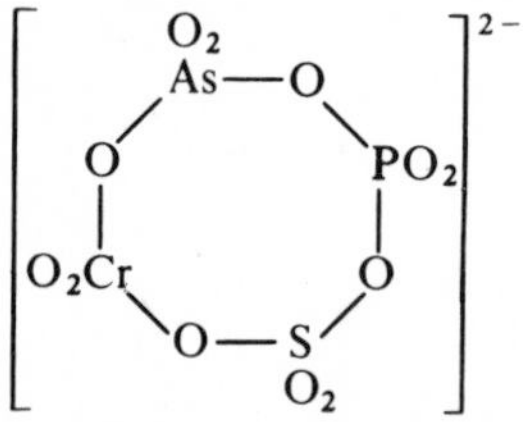

$$
\left[
\begin{array}{c}
O_2 \\
As\!-\!O \\
O \qquad PO_2 \\
O_2Cr \qquad O \\
O\!-\!S \\
O_2
\end{array}
\right]^{2-}
$$

cyclo-arsenatochromatosulfatophosphate(2−)

For a polyanion with an obvious central atom, the other components are named in alphabetical order as ligands, e.g.:

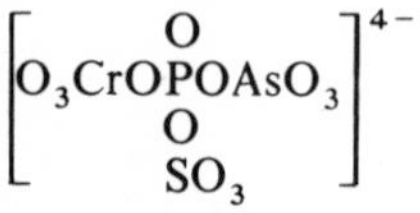

$$
\left[
\begin{array}{c}
O \\
O_3CrOPOAsO_3 \\
O \\
SO_3
\end{array}
\right]^{4-}
$$

(arsenato)(chromato)(sulfato)phosphate(4−)

Note the use of parentheses to indicate separate linkage to the central atom.

Condensed heteropolyanions are named by designating the octahedral atoms surrounding the neutral atoms by prefixes such as wolframo- (or tungsto-), molybdo-, etc. Examples are:

$[PW_{12}O_{40}]^{3-}$ dodecawolframophosphate(3−) *or* 12-wolframophosphate(3−)

$[PMo_{10}V_2O_{39}]^{3-}$ decamolybdodivanadophosphate(3−)

$Li_3H[SiW_{12}O_{40}]\cdot24H_2O$ trilithium hydrogen dodecawolframo-silicate-24-water

ADDITION COMPOUNDS

Molecular compounds, solvates, and clathrates are, according to the 1970 IUPAC rules, to be designated by connecting the names of the components by hyphens (or en rules) and stating the molar proportions by Arabic numerals separated by a solidus (/) and in parentheses after the name. Boron compounds and water are always cited last, in that order. Other molecules are cited in order of increasing number; any that occur in equal numbers are cited in alphabetical order. In formulas, dots at the midway position replace the hyphens. For example:

$Na_2CO_3\cdot10H_2O$ sodium carbonate–water (1/10) *or* sodium carbonate decahydrate

$NH_3 \cdot BF_3$	ammonia–boron trifluoride (1/1)
$BF_3 \cdot 2H_2O$	boron trifluoride–water (1/2)
$8CHCl_3 \cdot 16H_2S \cdot 136H_2O$	chloroform–hydrogen sulfide–water (8/16/136)

NONSTOICHIOMETRIC CRYSTALS

Crystalline phases not conforming to strict stoichiometry are discussed at some length in the 1970 IUPAC rules (Section 9), where designations for phases of variable composition (berthollides), vacant and interstitial sites (including Schottky and Frenkel defects), surface sites, and doped materials are dealt with. A symbol □ is used in many such cases.

AFFIXES IN INORGANIC NOMENCLATURE

The useful list in *Table 2.12* is reproduced from the IUPAC 1970 rules.

Table 2.12 AFFIXES USED IN INORGANIC NOMENCLATURE

MULTIPLYING AFFIXES	(*a*) mono, di, tri, tetra, penta, hexa, hepta, octa, nona (ennea), deca, undeca (hendeca), dodeca, *etc.*, used by direct joining without hyphens
	(*b*) bis, tris, tetrakis, pentakis, *etc.*, used by direct joining without hyphens, but usually with enclosing marks around each whole expression to which the prefix applies
STRUCTURAL AFFIXES	italicized and separated from the rest of the name by hyphens
antiprismo	eight atoms bound into a rectangular antiprism
asym	asymmetrical
catena	a chain structure; often used to designate linear polymeric substances
cis	two groups occupying adjacent positions; sometimes used in the sense of *fac*
closo	a cage or closed structure, especially a boron skeleton that is a polyhedron having all triangular faces
cyclo	a ring structure†
dodecahedro	eight atoms bound into a dodecahedron with triangular faces
fac	three groups occupying the corners of the same face of an octahedron

continued

Table 2.12 (*continued*)

hexahedro	eight atoms bound into a hexahedron (e.g., a cube)
hexaprismo	twelve atoms bound into a hexagonal prism
icosahedro	twelve atoms bound into a triangular icosahedron
mer	meridional; three groups on an octahedron in such a relationship that one is *cis* to the two others which are themselves *trans*
nido	a nest-like structure, especially a boron skeleton that is very close to a closed or *closo* structure
octahedro	six atoms bound into an octahedron
pentaprismo	ten atoms bound into a pentagonal prism
quadro	four atoms bound into a quadrangle (e.g. square)
sym	symmetrical
tetrahedro	four atoms bound into a tetrahedron
trans	two groups directly across a central atom from each other; i.e. in the polar positions on a sphere
triangulo	three atoms bound into a triangle
triprismo	six atoms bound into a triangular prism
η (eta or hapto)	signifies that the atom or group so designated bridges two or more centers of coordination
σ (sigma)	signifies that one atom of the group is attached to a metal

*These prefixes are often used also with words such as -valent and -dentate. IUPAC and others prefer to use Latin prefixes with Latin words, thus univalent, bivalent, tervalent, quadrivalent, quinquevalent, sexivalent and multivalent (rather than polyvalent); similarly with dentate.

†A prefix *cyclo-* here is used as a modifier indicating structure and hence is italicized. In organic nomenclature, cyclo- is considered to be part of the parent name since it changes the molecular formula and therefore is not italicized.

REFERENCES

1. *See J. Chem. Soc.*, 1404 (1940)
2. *IUPAC Compt. Rend. 17th Conf.*, 98–119 (1953)
3. *IUPAC Nomenclature of Inorganic Chemistry, 1957 Rules*, Butterworths, London (1959)
4. *IUPAC Compt. Rend. 23rd Conf.*, 181–187 (1965)
5. *IUPAC Nomenclature of Inorganic Chemistry*, 2nd Ed., *Definitive Rules* 1970, Butterworths, London (1971) [reprint from *Pure Appl. Chem.*, **28**, 1 (1971)]; now available from Pergamon Press, Oxford
6. FERNELIUS, W.C., *How To Name an Inorganic Substance*, Pergamon Press, Oxford (1977)
7. *IUPAC Information Bulletin, Appendix* No. 55 (1977)
8. CAHN, R.S., INGOLD, SIR C., and PRELOG, V., *Angew. Chem., Internat. Ed.*, **5**, 385, 511 (1966)
9. BROWN, M.F., COOK, B.R. and SLOAN, T.E., *Inorg. Chem.*, **14**, 1273 (1975)

3

Organic: General

Introduction

Nomenclature rules published by the International Union of Pure and Applied Chemistry (IUPAC) are accepted by chemists of most countries and thus form the international authority for organic as for inorganic chemistry*.

The IUPAC Organic Nomenclature Rules A, B, and C, 1969[2], which supersede earlier versions[3], cover most of general organic chemistry but hardly any of the specialized fields. Nomenclature of organic derivatives of phosphorus, arsenic, antimony, and bismuth, and of organometallic compounds other than coordination complexes (*see* pp. 17, 26) is at present (1978) officially specified only in tentative rules[4] issued jointly by the IUPAC Commissions on the Nomenclature of Organic and of Inorganic Chemistry. These are discussed in Chapter 9. A number of fields of major interest in both biochemistry and organic chemistry have been treated jointly by Commissions of IUPAC and IUB (International Union of Biochemistry), and several valuable sets of specialist rules have resulted (*see* Chapter 8).

However, anyone attempting to learn nomenclature by reading these rules must find them often complex, arbitrary, contradictory, or vague (because of alternatives). These bad features arose from a wish by IUPAC (and all chemists) to retain as much of customary nomenclature as is still useful, because large sections of that customary usage are antique (e.g., acids and their derivatives are named by a dualistic principle reminiscent of the earliest chemists, and amines are often named on the mid-nineteenth century theory of types), and because chemists are loth to abandon a host of abbreviated names. Add to these the many thousands

---

*Early attempts to reach agreed organic chemical nomenclature have been recorded by Verkade[1] in a series of papers containing much hitherto little-known or unpublished material.

of purely trivial names, some old, some new, and then a fully logical nomenclature, simple to understand and compatible with computer usage, might appear a desirable objective. Even if this were possible, however, such a completely new start would meet at most only gradual acceptance by chemists accustomed to current terminology. So until the omens are more favourable, current IUPAC nomenclature needs to be mastered in spite of its deficiencies, at least because its general principles are used to a large extent in *Chemical Abstracts (CA)* indexes. In the account that follows here an attempt is made to set out not so much the formal rules as the most important principles, together with indications as to where information on the more recondite matters can be found.

Chemical Abstracts has published substantial discussions of the basis of naming organic compounds, designed specifically for users of their chemical substance indexes. The 1972 edition[5] replaced and considerably modified the next earlier ones[6]; but there is to be no further significant change until at least 1982. A general guide in English to organic nomenclature, emphasizing principles and systematization, was edited by Fletcher, Dermer and Fox[7]; it presents the subject with fewer alternatives than the IUPAC rules permit, but in more detail than the present book. There also exist a number of smaller manuals for use of the beginner[8], and one for teachers, advocating strict adherence to systematic IUPAC names, is being widely adopted for both organic and inorganic chemistry[9].

It will be most convenient, if, before proceeding to the principles of nomenclature themselves, we first define certain technical terms and the conventions used in committing chemical names to paper.

Technical Terms

Systematic name

A name composed wholly of syllables specially coined or selected to describe structural features: hexane, thiazole.

Trivial name

A name no part of which is used in a systematic sense: xanthophyll, furan.

Semisystematic = semitrivial name

A name of which only a part is used in a systematic sense: meth*ane*, but*ene*, calcifer*ol*. Most names in organic chemistry belong to this class; many are often spoken of loosely and incorrectly as simply 'trivial'.

Parent

We may think of benzene as the chemical parent of nitrobenzene because the latter is made from the former; and benzene is the parent name of nitrobenzene because the latter is derived from the former by a prescribed variation. However, parent names do not always reflect chemical parentage: ethane is hardly the chemical parent of ethanol, though ethane is the parent name of ethanol. Often there is a chain of parentage — hexane, 3-methylhexane, 3-(chloromethyl)hexane — or multiple parentage — benzanthracene from benzene and anthracene. It is important to note that a parent name may be systematic (e.g., hexane), trivial (e.g., furan), or semitrivial (e.g., methane).

Group or radical

A neutral group of atoms, bound together in some way, may be common to a number of compounds; e.g., CH_3, CH_2, OH, NO_2, CO, CO_2H. Such groups are often called radicals in nomenclature parlance (cf. p. 22). As noted earlier (*see* p. 22), free radicals are always given the designation 'free' in nomenclature work.

Function, functional group

The terms 'function' and 'functional group' entered English chemical nomenclature in IUPAC rules by translation from the 1930 French version (*fonction, fonctionnel*) without definition. A functional group is a group of atoms defining the 'function' or mode of activity of a compound. An alcohol owes its alcoholic properties to the functional group —OH; here the functional group, hydroxyl, is the same as the chemical group of the same name. A ketone owes its ketonic properties to the oxygen atom that is doubly bound (to carbon); the ketonic function is O=(C) (without the carbon), and this is not the same as the ketonic group O=C<. Similarly the carboxylic function is **1**, and **2** is the carboxyl group. In chemical writing we often find 'function', 'group', and 'radical' used interchangeably; but they are not interchangeable. Usually group or radical expresses the meaning intended.

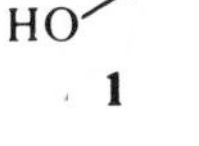

     1 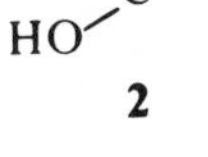   **2**

Incidentally, the division of substituents into 'functional' and 'non-functional' in Beilstein's *Handbuch* does not conform to this definition; e.g., the nitro group certainly imparts a mode of activity to nitro-benzene, but is considered nonfunctional in the *Handbuch*.

Principal group

It is generally useful, for two main reasons, to name the functional group as a suffix, or, if there is more than one kind of functional group in the compound, to select one kind for citation as suffix. The first reason is that this pinpoints an important chemical class to which the compound belongs and whose properties can be expected; the second reason is that the selected suffix is used by various devices to control several other choices, e.g., the choice of parent compound and numbering. The selection of the suffix is made according to precise rules. These are described in later sections of this book, but it may be noted here that these rules are not fully logical — that a number of functional groups are never cited as suffix and that unsaturation is subject to exceptional treatment.

General adoption of the term 'principal group' for the kind of group named as suffix would allow 'function' and 'functional group' to be used for their proper purposes, obviating a confusion that is still often encountered.

Locant

This useful word denotes the numeral or letter that indicates the position of an atom or group in a molecule.

Conventions

Conventions concerning the exact way in which chemical names should be written are an annoyance because so many of them are not emphasized in speech and because they vary from country to country and in some cases from journal to journal within one country, partly for linguistic reasons and partly from personal preferences. Some of these differences chemists take in their stride, but others must be assiduously learnt: for instance, it is important to know the difference between 1,2,5,6- and 1,2:5,6- and that italicized prefixes are not alpha-beticized in most chemical indexes, so that isobutylamine appears under I whereas *sec*-butylamine appears under B.

Somewhat distressing are the differences between *CA* and the (British) Chemical Society. In the first three editions of this book the British conventions were used. In the fourth and the present edition, since computer output from United Kingdom Chemical Information Service (UKCIS) and other organizations is based almost wholly on American compilations, and since American publications are much the more numerous, it has seemed advisable to use American conventions. Differences between the two national habits are noted where appropriate.

Numerals

A numeral is a locant when it indicates the position of a substituent or bond in a structure. Two or more locants denoting the positions of two or more identical substituents are separated by commas, as in 1,2,4,5-tetrabromocyclohexane. When two or more operations each require two (or occasionally more) locants, the pairs (or triplets, etc.) of numerals are separated by colons, as in 1,2:5,6-di-*O*-isopropylideneglucitol. There is no space after the comma or colon. Single numerals, or such sets of numerals, are joined by hyphens to the following and any preceding parts of the name; however, in a few special cases a single letter is also part of the locant and then follows the numeral directly, between it and the hyphen or comma, e.g., 3*a*- or (7b)-.

In bicyclo, tricyclo, etc., names (*see* p. 86) and in certain spiro names (*see* p. 88) numerals are separated by full stops (periods), again without spaces; but such numerals are not locants.

Series of numerals are arranged in ascending order. An unprimed locant is lower than the same locant primed, a singly primed lower than a doubly primed, and any primed locant lower than a higher unprimed locant; e.g., an order would be 2,2,2′,3,3′,3″,4. Special significance attaches to departure from this order in specific cases, e.g., bicyclo and spiro names.

Locants are placed as early in a name as does not cause confusion. Simple examples are 2-chlorohexane, 2-chloro-3-methylhexane, and 2,3-dichlorohexane.

American practice extends this to locants for suffixes in cases such as 2-hexanol, 2,3-hexanediol, 2-phenanthrenecarboxylic acid, and 2-hexene. But this can be done only for one *type* of such ending; if there is more than one, the locant appearing first in the name is placed on the left and the others directly precede their suffixes, e.g., 2-hexen-4-yne, 3-hexen-5-yn-2-ol. That practice is followed in this edition. British custom is to place the locant always immediately in front of its suffix, as in hex-2-ene and hex-3-en-5-yn-2-ol, but most chemists in other countries dislike this as splitting spoken words unnecessarily.

There are also cases where no locant for a suffix can be moved to the left, e.g., when some other locant must by a special rule be attached at that place; examples such as 5a-cholestan-3-one, 2H-pyran-3-ol, and bicyclo[3.3.0]oct-2-ene illustrate this.

After all this, it should be noted that quite different rules apply to positions and punctuation of locants in many non-English languages (cf. pp. 50–51).

Letter locants

Single-letter locants are also common, often italicized atomic symbols; if for identical groups, such locants are arranged in alphabetical order, Latin before Greek, whether capital or lower case. Arrangement for primed letters follows the same principles as for primed numerals. Mixtures such as N,a,2-trimethyl occasionally occur.

The familiar o- (ortho), m- (meta), and p- (para) locants for benzene ring positions have been abandoned by CA in favour of numerals, e.g., 1,4-dichlorobenzene.

Multiplying prefixes

The normal multiplying prefixes used in organic chemistry are listed in Table 3.1.

Multiplicative prefixes bis-, tris-, tetrakis-, etc. are used in CA for all complex expressions. The commonest is for compound radicals,

Table 3.1 MULTIPLYING PREFIXES NORMALLY USED IN ORGANIC CHEMISTRY

1	mono	11	undeca*		
2	di	12	dodeca		
3	tri	13	trideca	30	triaconta
4	tetra			31	hentriaconta*
5	penta			32	dotriaconta
6	hexa	20	eicosa†		
7	hepta	21	heneicosa*	40	tetraconta
8	octa	22	docosa		
9	nona*	23	tricosa	100	hecta
10	deca	24	tetracosa	132	dotriacontahecta

*nona and un are from Latin; ennea and hen from Greek.
†The spelling icosa is recommended by IUPAC for inorganic chemistry.

e.g., bis(dimethylamino), tris(2-chloroethyl), etc.; and this is extended
to other situations where parentheses aid clarity, as in bis(diazo),
tris(hydrogenphthalate), and tris(methylene) (to avoid confusion with
trimethylene $-CH_2CH_2CH_2-$).

British practice has been more permissive. Bis-, etc., have been used
when the next syllable was numerical, as in bis-2,2,2-trichloroethyl,
parentheses being considered unnecessary; and di-(2-chloroethyl) has
been held in Britain to be as unambiguous with the parentheses as with
a bis prefix.

It will also be found that the first hyphen that in Britain would be
as in tris-(2-chloroethyl) and the like is considered in America to be
unnecessary when a parenthesis follows; the American custom would
be tris(2-chloroethyl).

Parentheses, brackets, etc.

As just noted, parentheses ( ) are used in *CA* around all compound
radical names to avoid any possible confusion, e.g., 2-(bromomethyl),
(phenylazo), and even 2-(2-thienyl), etc. In more complex cases, square
brackets [ ] are added, e.g. 2-[1-(diethylamino)ethyl]benzoic acid, or
9-[2-(dichloromethoxy)ethyl]phenanthrene. British practice has been
less rigid, allowing chemical sense to dictate when parentheses or indeed
also braces { } are needed. Chemical sense is also used in American
practice when ambiguity might otherwise arise: for instance, CH_3PHCl
is chloromethylphosphine, but PH_2CH_2Cl is (chloromethyl)phosphine;
so also glucose 6-(dihydrogen phosphate); and abbreviated names such
as methoxy are not placed within parentheses whereas unabbreviated
(heptyloxy, etc.) are. Sometimes the separateness of ligands is emphasized with parentheses, as in chloro(methyl)mercury, CH_3HgCl.

Many other special uses for enclosing marks will be found in the
following pages.

Italics

Since italicized syllables are not 'counted' in alphabetization, their
use in the following pages should be noted. The purpose of italics is
almost always simply to ensure that index entries occur in the most useful places, keeping isomers and related compounds in suitable groups.
This being the object, the fact that no difference is made in speech
becomes irrelevant.

A lower-case italic prefix at the beginning of a line or after a full
stop (period) is not given an initial capital — the next letter becomes a

capital, as in *m*-Xylene or *trans*-2-Butene. Conversely, a capitalized italic prefix does not serve as the capital letter marking the beginning of a sentence, e.g., *N*-Methylacetanilide.

Elision of vowels

Instructions for elision or retention of vowels are given in the IUPAC rules (ref. 2, pp. 83–84) which should be consulted for details. The following are common cases that occur in *CA* and much general practice.

(1)   Before a functional suffix that begins with a vowel, e.g., 2-hexanone (not 2-hexaneone).
(2)   Between 'a' of a numerical prefix and a functional suffix that begins with a vowel, e.g., benzenehexol (not benzenehexaol).
(3)   In Hantzsch–Widman names (*see* p. 90), e.g., oxazole (not oxaazole).
(4)   When a prefix such as cyclopropa-, benzo-, or naphtho- is attached to a further part of a ring name that begins with a vowel, e.g., cyclobutindene.

Note that (2) is not applied in inorganic chemistry by IUPAC (*see* p. 11).

Addition of vowels

An 'o' is occasionally inserted for euphony between a consonant ending one part of a name and a consonant (notably 'h') beginning the next part, e.g., acetohydroxamic acid (not acethydroxamic).

Formulas

In simple groups the central atom is written first, e.g., CH_3, CH_2, NO_2, $N(CH_3)_2$, $NH(CH_3)$, etc.; CH_3CH_2 may be shortened to C_2H_5, and similarly for some other alkyl groups. When such a group appears on the left of a formula the order may be reversed so as to emphasize the bonded atoms, as in $O_2N-CH_2CH_2-OH$ or $ON-O-CH_2CH_2OH$.

In British journals successive groups in a linear formula have often been separated by points at the midway position, as in $CH_3 \cdot CH_2 \cdot CH_2 \cdot OH$. But it is widely held elsewhere that such points should be restricted to denoting free radicals or addition compounds. Such 'British' midway points are never used in American practice in linear organic formulas; to emphasize the existence of a bond, a hyphen or 'rule' is inserted, as in $O_2N-CH_2-CH_2OH$.

Parentheses are used in three ways in linear formulas. In one mode they enclose groups identically bonded to a neighbouring atom, as in $(CH_3)_2CHC(CH_3)_3$. Another meaning is conveyed in $Cl(CH_2)_4CHO$, where the enclosed groups are understood to be consecutively bound in a chain. Finally, parentheses may mean that the group they mark is *not* part of the chain, but is a side group or branch, e.g., in $CH_3CH_2CH(CH_3)$-CH_2OH or $Cl_2P(O)OCH_3$. *Chemical Abstracts* sets off all lateral groups in this way, as in $CH_3CH(OH)CH_2OH$ and $C_6H_5CH(NH_2)CH_2CH_3$, but misreading such formulas is hardly possible even if the parentheses are omitted.

Certain group symbols have no official IUPAC sanction but are very frequent in British and American publications, although usage in these varies somewhat. The commonest are shown in *Table 3.2*.

Table 3.2 SYMBOLS FOR SIMPLE ORGANIC SUBSTITUENT GROUPS

	American symbol	British symbol		American symbol	British symbol
CH_3-	Me	Me	$CH_3CH_2CH(CH_3)-$	*sec*-Bu	Bu^s
CH_3CH_2-	Et	Et	$(CH_3)_3C-$	*t*-Bu	Bu^t
$CH_3CH_2CH_2-$	Pr	Pr^n	$CH_3C(O)-$	Ac	Ac
$(CH_3)_2CH-$	*i*-Pr	Pr^i	C_6H_5	Ph	Ph
$CH_3CH_2CH_2CH_2-$	Bu	Bu^n	$C_6H_5C(O)-$	PhCO	Bz
$(CH_3)_2CHCH_2-$	*i*-Bu	Bu^i		or Bz	

Straight-chain alkyl groups are named without '*n*-', e.g., heptyl, but the formulas should be written as e.g., *n*-C_7H_{15}, not C_7H_{15} which is a generic formula.

Some generic symbols are widely used. Chief among these is R, usually to represent a hydrocarbon radical, and preferably a univalent one*. A series of such should be written R, R', R'' ... R^1, R^2, R^3 ...., but not R_1, R_2, R_3 (which of course denote 1, 2, or 3 identical R groups). Ar is sometimes employed to represent aryl groups in general, in spite of its identity with the atomic symbol for argon, and X for a group that needs to be differentiated from the R's.

In some specialist nomenclature, particularly in biochemistry and for coordination compounds, special symbols are essential for brevity; acronyms are often used. Thus in reference to common organic solvents, THF often stands for tetrahydrofuran, and DMSO for dimethyl

*Univalent versus monovalent: *see* footnote to Table 2.12.

sulfoxide. It is essential for good communication that official lists of accepted abbreviations be published, and that any other symbols used by authors be defined in the text.

The old problem of representing benzene by formula is met in IUPAC rules, in *CA*, and in most British chemical literature by using **3** or **4**, but American textbooks and journals have lately preferred **5** or **6**.

Hyphenation of chemical words

Hyphenation of chemical terms is much less common in American than in British practice. In the former, generic names of difunctional compounds are written without hyphens unless they are full substitutive names (*see* p. 55), when they are single words. Thus chloro alcohol and amino acid are correct, but so are bromoalkanols and aminoalkanoic acids (not aminocarboxylic acids). The British usually add hyphens in such names, e.g., chloro-alcohol and oxo-steroid, as well as in two-part names in which the first part ends in a voiced vowel or '-y', e.g., amino-derivative, thia-compound, methoxy-group (but methyl substituent, etc.). The hyphen is also used in Britain to separate two identical letters in a chemical name (e.g., tetra-acetate and methyl-lithium) and to help in reading such words as co-ordination and un-ionized. None of these hyphens is considered necessary in America.

Other Languages

By and large, chemical names are much easier than current text to translate from foreign languages, for they almost always bear at least some resemblance to one or other of the not too ancient systems current in the UK or USA. It is the long-known compounds that are the usual exceptions to this emollient statement, for example, the common metals, the common gases, and a few organic compounds such as acetic and formic acid. Most troublesome are the resemblances that hide differences, and the following notes may be useful.

In German the main trap is that a terminal '-e' on a word denotes a plural (with few exceptions); 'Anilin' is aniline, 'Nitroaniline' are nitro-anilines; 'Phenol' is phenol, 'Phenole' are phenols; thiazole becomes 'Thiazol', and thiazoles become 'Thiazole'. Thus the distinction of basic

'-ine' from nonspecific '-in' (tannin, protein, anthocyanin, vitamin, etc.) and of alcoholic or phenolic '-ol' from '-ole' (denoting a five-membered aromatic ring) cannot be used in German, and the true significance of the -e in German may be overlooked by a hasty reader.

The ability, in German, to pile nouns on one another shines also through their nomenclature, but, for example, 'Acetessigsäureethyl-esterdinitrophenylhydrazon' merely requires careful dissection. More subtle are the results where in German there may be more hyphens than in English, as in 'Chloro-bromo-phenylnaphtalin', or fewer hyphens as in 'Chloro- und Bromonaphtaline'. Perhaps special note may be needed that the German words for the simplest aromatic hydrocarbons are Benzol, Toluol, and Xylol, and have persisted until recently in British–American technology.

In general, the influence of Beilstein/Stelzner nomenclature in German literature is still profound, though IUPAC nomenclature is steadily gaining ground, principally because of its greater simplicity (or lesser complexity!) and wider applicability together with the many publications of the Chemical Abstracts Service.

The problem of terminal '-e' has a different result in French, viz., that, for phonetic reasons, '-in' and '-ane' endings for non-nitrogenous rings (*see* p. 95) become '-inne' and '-anne'. But the features in French nomenclature that most strike the English-speaking reader are that locants follow the names of the structural features concerned, and the whole name is divided into small words, as in:

$$ClCH_2-CH-CH_2-CH-CH_2\,CH_2\,CH_3$$
$$CH_2\,Cl \qquad CHClCH_3$$

Chloro-1 (chloro-1 éthyl)-4 chlorométhyl-2 heptane

There is also the formation of salt and ester names as, for example, 'acetate de sodium' and 'malonate de diéthyle'.

An excellent exposition of French organic nomenclature, reproducing recent IUPAC rules (with occasional modifications) and including a brief historical introduction and very valuable comments on individual sections and rules, has been published by Lozac'h[10]; it is a much more sophisticated and detailed treatment than is given in the present book.

A German-language version of IUPAC rules is being published in sections as a loose-leaf work[11]. Holland[12] has written, in German, a scholarly account of organic nomenclature, both IUPAC principles and other usage, current and past; some 916 references are cited.

The Russian chemical literature now presents fewer problems than it used to, because some of it is also published in English translations;

but the reader should be warned that Russian nomenclature seems to be left to the authors' whims and is often modelled on Beilstein rather than IUPAC – a feature that can be particularly troublesome in their numerous and extensive organophosphorus papers.

In other countries, notably Scandinavia, Japan, Holland and Belgium, the tendency to publish in English increases continuously, then mainly in the style of *CA* and IUPAC.

REFERENCES

1. VERKADE, P.E., *Bull. Soc. Chim. France*, 1807 (1966); 4009 (1967); 1358 (1968); 3877, 4297 (1969); 2739 (1970); 1634, 4299 (1971); 1961 (1973); 555, 1119, 2029 (1975); 1445 (1976); 457 (1977); 13 (1978)
2. *IUPAC Nomenclature of Organic Chemistry. Definitive Rules for: Section A. Hydrocarbons; Section B. Fundamental Heterocyclic Systems; Section C. Characteristic Groups Containing Carbon, Hydrogen, Oxygen, Nitrogen, Halogen, Sulfur, Selenium and/or Tellurium*, 1969. A, B 3rd Ed.; C 2nd Ed., Butterworths, London (1971); now available from Pergamon Press, Oxford
3. Ref. 2: *Sections A and B*, 1st Ed., 1958; 2nd Ed., 1966. *Section C*, 1st Ed., 1965 [also printed in *Pure Appl. Chem.,* **11**, 1 (1965)]
4. *IUPAC Information Bulletin, Appendix* No. 31 (1973)
5. *Chemical Abstracts, Index Guide to Volume 76*, p. 211–1011 (1972); *Ninth Collective Index, Vols. 76–85* (1972–1976), *Index Guide*, paragraphs 101–212
6. *Chemical Abstracts, Subject Index – Introduction,* **56**, 11N–98N (1962); **66**, 1I–40I (1967)
7. FLETCHER, J.H., DERMER, O.C. and FOX, R.B. eds., *Nomenclature of Organic Compounds: Principles and Practice* (*Advances in Chemistry Series No.* 126), American Chemical Society, Washington, D.C. (1974)
8. TINLEY, E.H., *Naming Organic Compounds: A Guide to the Nomenclature Used in Organic Chemistry*, Alchemist Publications, London (1962), 48 pp.; RUNQUIST, O.A., *Programmed Review of Organic Chemistry: Nomenclature,* Burgess Publishing Co., Minneapolis (1965), 85 pp.; BANKS, J.E., *Naming Organic Compounds*, 2nd Ed., Saunders, Philadelphia (1976), 309 pp.; TRAYNHAM, J.G., *Organic Nomenclature: A Programmed Introduction*, Prentice-Hall, Englewood Cliffs, N.J. (1966), 129 pp.; JOHNSON, C.R., *Organic Nomenclature*, Worth Publishers, New York (1976), 130 pp.
9. THE ASSOCIATION FOR SCIENCE EDUCATION, *Chemical Nomenclature, Symbols and Terminology*, College Lane, Hatfield, Herts. (1972), 66 pp. Boards in charge of school examinations in the United Kingdom and elsewhere are now restricting nomenclature to the forms described in this book, so that pupils leaving school will increasingly be familiar with IUPAC terms and will recognise, for example, ethanol but not ethyl alcohol, and butanedioic acid but not succinic acid. A revised edition of this publication is expected in 1979.
10. LOZAC'H, N., *La Nomenclature en Chimie Organique*, Masson et Cie., Paris (1967) (Vol. 6 of *Collection de Monographies de Chimie Organique, Compléments au Traité de Chimie Organique*, under the direction of A. Kirrmann, M.-M. Janot, and G. Ourisson)

11. DEUTSCHEN ZENTRALAUSSCHUSSES FÜR CHEMIE, *Internationale Regeln für die chemische Nomenklatur und Terminologie*, Verlag Chemie, Weinheim (1975)
12. HOLLAND, W., *Die Nomenklatur in der organischen Chemie*, 2nd Ed., Verlag H. Deutsch, Zürich (1973)

4

Organic: The Principles

Types of Nomenclature

There is a fundamental distinction between the use of trivial and of
systematic names: trivial names refer to compounds, systematic names
to structures, i.e., structural formulas. Trivial names are independent of
structure; they can be, and often are, assigned before the structure is
known; and when the structure is known, the one name embraces all
dynamic variations due to tautomerism, etc. A systematic name, being
derived from one formula, cannot, if it is accurately descriptive, apply
to a tautomer thereof (though normally it covers hybrid structures
representing resonance, hyperconjugation, etc.). Of the approximately
four million organic compounds at present recorded, many thousands
have trivial or semitrivial names, each of which, not being wholly
logical, requires some feat of memory for recall of the relevant structure.
No one, of course, remembers more than a tiny proportion of them;
yet trivial names are inevitable when of long tradition or when the
systematic name is too unwieldy for use — certainly for macromolecules
such as proteins and nucleic acids. The important thing is to avoid
coining new trivial names just for the fun of it and in simple cases grad-
ually to replace the old by the systematic.

In a semitrivial name an indication of partial structure is, by
definition, built in (e.g., the alcoholic group in ethanol); it is essential
that this be done correctly, e.g., that acidic phenols be not *named*
acids (picric 'acid' appears too well entrenched to dislodge, but names
such as carbolic and cresylic are obsolete). Any modifications of trivial
names to describe derivatives should be made in accordance with the
rules for systematic nomenclature.

Owing to the formation, by covalent linkages, of chains and rings
of carbon atoms, alone or with heteroatoms (i.e., atoms other than
carbon), the structures of organic compounds differ fundamentally

from those of many inorganic compounds. The two nomenclatures therefore use quite different methods. Organic nomenclature has, however, been of slow growth and in its present state no fewer than nine general principles and several specialized principles can be distinguished. The latter will be recorded at appropriate places. The former can be outlined at once as follows; in most cases they apply equally to trivial, semitrivial, and systematic names.

(1)   The basic principle is substitution, the replacement of hydrogen by an atom or group, e.g., of H by Cl (chlorination), by NO_2 (nitration), by CH_3 (methylation), even though this replacement may not be the method of synthesis.

Some restriction is necessary. Chlorination, for instance, is only replacement of H by Cl; the term should not be used for addition, as in:

$$C_6H_5CH=CH_2 + Cl_2 \rightarrow C_6H_5CHClCH_2Cl$$

or for the reaction

$$\geqslant C-OH \rightarrow \geqslant C-Cl$$

Hydrogenation is an obvious exception.

Substitution is indicated by a suffix (ethane, ethanol) or a prefix (benzene, chlorobenzene), the loss of hydrogen not being stated.

(2)   By substitutive nomenclature one reaches the name 2-naphthylethanol for $2-C_{10}H_7CH_2CH_2OH$, indicating substitution of the naphthyl radical into ethanol. Such names have, however, been considered poor for indexing; the parent is ethanol, and after inversion the index entry would be Ethanol, 2-naphthyl-; but the compound is more suitably indexed under the larger part, naphthalene, and this can be achieved by juxtaposing the two names naphthalene and ethanol and adding the locant numeral as prefix. This yields 2-naphthaleneethanol (for indexing under N), the loss of *two* hydrogen atoms being assumed in this process (just as the loss of *one* is assumed in substitutive names). This is termed 'conjunctive nomenclature'. Further detail is given on pp. 68–69.

(3)   What is superficially a mixture of substitutive and conjunctive nomenclature is the (universal) practice with prefixes for bivalent groups, such as $-O-$ oxy-, $>C=O$ carbonyl-, and $-S-$ thio-. Suppose we wish to treat CH_3O_2C- as a substituent: we place the radical name methoxy in front of the bivalent radical name carbonyl and obtain methoxycarbonyl CH_3O-CO-, *no* hydrogen being lost in the process; then we can substitute this group into, say, glycine $NH_2CH_2CO_2H$, with the usual loss of one hydrogen atom in the process, and in this

way we obtain N-(methoxycarbonyl)glycine $CH_3O_2CNHCH_2CO_2H$.

(4)   For some classes, old names indicating the function of the compounds survive, e.g., ethyl alcohol, diethyl ether, ethyl methyl ketone, acetic acid. The systematist would like these to disappear, for (except for acids) there are alternatives more in keeping with current practice; but such names die hard, if only because the chemist is, after all, interested mainly in the functions, i.e., modes of activity, of his compounds. This system has been termed radicofunctional nomenclature by IUPAC, because the functional class name (alcohol, ketone, etc.) is preceded by a radical name or names (ethyl, acetic, etc.).

(5)   Replacement nomenclature, also called 'a' nomenclature: sometimes replacement of a heteroatom in a compound A by carbon would yield a substance B that is more readily recognized or more simply named. It is then possible to name the compound B and indicate the presence of the heteroatom by a prefix ending in 'a'. Thus, pyridine might be termed azabenzene. Actually this method is normally reserved so as to take advantage of a useful trivial name for a parent cyclic compound or for compounds containing a multiplicity of heteroatoms. Its extension to aliphatic compounds has been described by IUPAC[1a] and by CA[2] in different ways, but both restrict its application to relatively complex compounds so as to avoid creation of unfamiliar and unnecessary expressions such as 3-oxapentane for $CH_3CH_2OCH_2CH_3$. Authors having cause to name aliphatic chains containing many heteroatoms should consult the sources cited.

(6)   A few additive reactions survive as a basis for names, e.g., styrene oxide. This is, in fact, one of the few additive names that are worth preserving, for the alternatives 1,2-epoxyethylbenzene and 2-phenyloxirane are less easily recognized by some chemists. Names such as ergosterol dibromide are also valuable as trivial names preserving the parent component and, above all, its stereochemistry. In general, however, additive nomenclature should be avoided: it forms no part of modern nomenclature. Of course, there is an exception: hydro for addition of hydrogen!

   A different kind of additive nomenclature is used when an element increases its valence: e.g., pyridine gives pyridine 1-oxide.

(7)   Subtraction, i.e., removal of atoms, is indicated in a few cases. The most abundant examples of this are the '-ene' and '-yne' suffixes used to designate the presence of carbon–carbon double and triple bonds, respectively (i.e., loss of pairs of hydrogen atoms). Rarely such loss is cited by the prefix didehydro (loss of 2H) or the Greek letter Δ. Other subtractive prefixes are 'anhydro-' (loss of H_2O), 'nor-' (loss of CH_2), and 'de-' (CA) or 'des-' (IUPAC/IUB) (loss of a specified group: e.g. de-N-methyl).

(8)   Many cyclic skeletons have trivial names; complex cases are

handled by joining names of simpler components, by methods described in Chapter 5. There are also systems for such compounds where specific syllables have prescribed meanings for ring structure.

(9)   A few chemical operations can be designated by specific affixes, e.g., '-lactone', 'seco-', '-oside'.

None of these methods alone suffices for the whole range of chemistry; some are partial alternatives, e.g., (1)/(2), (1)/(4), and (2)/(4); some have severely restricted use. Much the commonest is simple substitutive nomenclature.

Also a warning is perhaps appropriate already here that the special treatments of acids and their derivatives (*see* pp. 107–112), aldehydes (*see* p. 115), and amines (*see* pp. 120–121) do not fit wholly into any of the classes named above.

The Approach to a Name

With such an array of possible procedures how does one set about naming a compound of known structure? Occasionally the chemistry under discussion is such that the expert may be justified in breaking the rules: but here let us forget such relatively rare occasions and confine ourselves to the classical methods and first to the preferred substitutive nomenclature. Then perhaps the obvious way is to look for a large recognizable unit − a ring structure or a long chain. Nevertheless, that would be wrong. The first thing to do is to seek out the substituent groups: OH, NH_2, CO_2H, SO_3H, OCH_3, $CO_2C_2H_5$, etc.; then from those present, the 'senior', so-called 'principal', group is found by means of *Table 4.2* (*see* p. 60); it is important to realize that use of this Table applies to choice of principal group no matter which form of nomenclature is used. That group, and that alone, is cited as suffix; any others become prefixes. The principal group sets the whole pattern of nomenclature and numbering, and it is vital to fix that group before anything else is done. Two simple examples show this: $C_{10}H_7C_6H_4CO_2H$ is a naphthylbenzoic acid and not a (carboxyphenyl)naphthalene, whereas $C_6H_5C_{10}H_6CO_2H$ is a phenylnaphthoic acid; $NH_2CH_2CH_2OH$ is 2-aminoethanol, and not 2-hydroxyethylamine (because OH is senior to NH_2).

For the rest of the name one works back from the principal group. The possibilities are so various that the following outline will appear confusing, but read slowly it will be found to make sense.

(1)   Suppose that there is only one type of principal group and that it is attached to an aliphatic chain. If there is also only one group of this

type, choose the most unsaturated chain to which it is attached*. If there are several groups of the same type, i.e., more than one principal group, attached to an aliphatic chain, then choose the straight chain bearing the most of them; if there are two such chains, choose the more unsaturated, then the longer*, as in the previous case. Next number the chain, with due regard to unsaturation and substituents, but giving the lowest available number(s) to the carbon atom(s) bearing the principal group(s), as explained later. Lastly, name the chain, add the suffix for the functional group(s), name the other substituents as prefixes, and add them in alphabetical order to complete the name. Further complications are described in succeeding pages.

(2)   Alternatively, suppose the principal group is attached to a ring. Then name the cyclic system, as explained later; number it, giving the lowest available numbers first to heteroatoms, then to carbon atoms bearing 'indicated' hydrogen (see p. 82) if any, then to atoms bearing the principal group, and finally to those atoms linked to the other atoms and groups to be cited; and write down the name.

(3)   If the principal group is attached to a chain and that in turn to a ring, an author may treat the cyclic radical as a substituent into the chain; or conjunctive nomenclature (see p. 68) may be used.

(4)   A variant of these procedures is useful for compounds with the symmetrical structural pattern $X-Y-X$ when each of the units X contains the same principal group for citation as suffix. Examples are:

$$p\text{-}HO_2CCH_2-C_6H_4-CH_2CO_2H \text{ and}$$
$$H_2NCH_2CH_2-O-CH_2CH_2NH_2$$

The principle is to cite the X groups with a prefix 'di', and to place in front of this the name of the bivalent group Y (i.e., phenylene and oxy in the two cases exemplified). This method is discussed in some detail on p. 73.

There are only a few steps in each case (1)–(4) and, if taken in the right order, they lead simply to the answer. Even though these are simplified directions and more complex situations are frequent, it is surprising how many of the compounds met in ordinary chemistry can be named merely by these simplified procedures. Of course, it is easier to write 'name the ring' than to do it, and tricky problems are often met. So in the following pages the procedures given will be expanded, and some of the problems discussed, but to reach them we must start again systematically at the beginning.

*CA now prefers the longest chain before considering unsaturation in selection of a parent name.

The Principal Group

Approaching a name again from the beginning, we must staft with what
we have just said is the first thing to do — selecting the principal group,
i.e., selecting the type of group to be named as suffix.

But we must here at once note that there may not be any principal
group in the compound; that will be the case if our compound is a
hydrocarbon or heterocycle or has as substituents only groups that
must be named as prefixes. The complex matter of naming hydrocarbons
and heterocycles will require separate treatment (*see* Chapter 5). For
the moment we shall remain with the groups.

As to groups that cannot be named as suffix, we find that there are
fewer of these in substitutive than in the older radicofunctional nomenclature and, since substitutive nomenclature is much the more widely
used as well as being the more modern, we shall begin with that.

PRINCIPAL GROUPS IN SUBSTITUTIVE NOMENCLATURE

Table 4.1 lists common groups that, in substitutive nomenclature,
can be named only as prefixes, together with the names of those
prefixes.

Table 4.1 SOME GROUPS CITED ONLY AS PREFIXES IN SUBSTITUTIVE
NOMENCLATURE

Group	Prefix	Group	Prefix
$-Br$	bromo	$=N_2$	diazo
$-Cl$	chloro	$-N_3$	azido
$-ClO$	chlorosyl	$-NO$	nitroso
$-ClO_2$	chloryl	$-NO_2$	nitro
$-ClO_3$	perchloryl	$-N(O)OH$	*aci*-nitro
$-F$	fluoro	$-OH$	R-oxy
$-I$	iodo	$-SR$	R-thio (similarly R-seleno
$-IO$	iodosyl		and R-telluro
$-IO_2$	iodyl (replacing		Hydrocarbon or heterocyclic
	iodoxy)		substituents
$-I(OH)_2$	dihydroxyiodo		
$-IX_2$	X may be halogen		
	or a radical, and		
	the prefix names		
	are dihalogenoiodo,		
	etc., or, for radicals,		
	patterned on		
	diacetoxyiodo		

Table 4.2 SOME GENERAL CLASSES OF COMPOUND IN THE ORDER IN WHICH THE RELEVANT GROUPS HAVE DECREASING PRIORITY FOR CITATION AS PRINCIPAL GROUP

1.	'Onium and similar cations
2.	Acids: in the order CO_2H, $C(=O)OH$, then successively their S and Se derivatives, followed by sulfonic, sulfinic acids, etc.
3.	Derivatives of acids: in the order anhydrides, esters, acyl halides, amides, hydrazides, imides, amidines, etc.
4.	Nitriles (cyanides), then isocyanides
5.	Aldehydes, then successively their S and Se analogues; then their derivatives
6.	Ketones, then their analogues and derivatives, in the same order as for aldehydes
7.	Alcohols, then phenols; then S and Se analogues of alcohols; then esters of alcohols with inorganic acids*; then similar derivatives of phenols in the same order
8.	Amines; then imines, hydrazines, etc.

*Except esters of hydrogen halides (*see Table 4.1*).

Table 4.2 lists, as classes, a greater number of groups that *can* be named as suffixes; any one of these, as sole substituent or accompanied only by group(s) listed in *Table 4.1*, is treated as the principal group, i.e., *must* be named as suffix in substitutive nomenclature or as functional class name (*see* following section) in radicofunctional nomenclature. Examples are hexanoic acid, pentanal, 3-acenaphthenecarboxylic acid, 1-chloro-2-butanol, and 4-methoxycyclohexanone. If two or more identical groups that qualify as principal group are present in the molecule, these are named with prefixes di-, tri-, etc., as in 2,3-hexanediol, 4-methoxy-1,3-cyclohexanediol, or 1,3,5-pentanetricarboxylic acid.

Frequently a compound contains more than one type of group listed in *Table 4.2*, and since it is a rule that only one type can be named as suffix an order of priority is required; that order is the order in which the classes are listed in *Table 4.2*. The order appears arbitrary but it is based on a study made many years ago by *CA* of the majority usage by chemists at that time when no official order existed.

Of course, any substituents from this list that are present in the compound and not selected as principal group must be named as prefixes; so *Table 4.3* lists the names of the most important of these groups as both suffix and prefix (*see* p. 61). Simple examples are 2-aminoethanol, *o*-aminophenol, 2-oxocyclopentanecarboxylic acid, and 3-ethoxy-4-fluorobenzamide.

Table 4.3, copied from the IUPAC rules, contains in the column headed *Suffix* various two-word entries for acids and some of their derivatives, and four blanks near the foot of the column. For the former

Table 4.3 IUPAC SUFFIXES AND PREFIXES FOR SOME IMPORTANT GROUPS IN SUBSTITUTIVE NOMENCLATURE

Class	Formula*	Prefix†	Suffix‡
Cations		-onio-	-onium
		-onia-	–
Carboxylic	$-CO_2H$	carboxy	-carboxylic acid
	$-(C)(=O),OH$	–	-oic acid
Sulfonic acid	$-SO_3H$	sulfo	-sulfonic acid
Salts	$-CO_2M$	–	metal . . . carboxylate
	$-(C)(=O), OM$	–	metal . . . oate
Esters	$-CO_2R$	R-oxycarbonyl	R . . . carboxylate
	$-(C)(=O),OR$	–	R . . . oate
Acid halides	$-CO-$Halogen	haloformyl	-carbonyl halide
	$-(C)(=O)$,halogen	–	-oyl halide
Amides	$-CO-NH_2$	carbamoyl	-carboxamide
	$-(C)(=O),NH_2$	–	-amide
Amidines	$-C(=NH)-NH_2$	amidino	-carboxamidine
	$-(C)(=NH),NH_2$	–	-amidine
Nitriles	$-C\equiv N$	cyano	-carbonitrile
	$-(C)\equiv N$	–	-nitrile
Aldehydes	$-CHO$	formyl	-carbaldehyde
	$-(C)H,(=O)$	oxo	-al
Ketones	$>(C)=O$	oxo	-one
Alcohols	$-OH$	hydroxy	-ol
Phenols	$-OH$	hydroxy	-ol
Thiols	$-SH$	mercapto	-thiol
Hydroperoxides	$-O-OH$	hydroperoxy	–
Amines	$-NH_2$	amino	-amine
Imines	$=NH$	imino	-imine
Esters	$-OR$	R-oxy	–
Sulfides	$-SR$	R-thio	–
Peroxides	$-O-OR$	R-dioxy	–

*Carbon atoms enclosed in parentheses are included in the name of the parent compound and not in the suffix or prefix.
†*Chemical Abstracts* has recently changed haloformyl, carbamoyl, and amidino to halocarbonyl, aminocarbonyl, and aminoiminocarbonyl, respectively.
‡*Chemical Abstracts* has recently changed carboxamidine, amidine, and carbaldehyde to carboximidamide, imidamide, and carboxaldehyde, respectively.

see pp. 107–112; the latter groups can be named only with prefixes in substitutive nomenclature.

PRINCIPAL GROUPS IN RADICOFUNCTIONAL NOMENCLATURE

In radicofunctional nomenclature we are dealing with names that consist of two or three words, the last word stating the function and the other(s) specifying the rest of the molecule in radical form. *Table 4.4*

Table 4.4 SOME FUNCTIONAL CLASS NAMES USED IN RADICO-
FUNCTIONAL NOMENCLATURE, IN ORDER OF DECREASING PRIORITY
FOR CHOICE AS SUCH

X in acid derivatives
    $RCO-X$, RSO_2-X, etc.
	name of X; in the order hydroxide, fluoride,
	    chloride, bromide, iodide, cyanide, azide,
	    etc.; then their S, followed by their Se
	    analogues

$-CN$, $-NC$	cyanide, isocyanide
$>CO$	ketone, then S, then Se analogues
$-OH$	alcohol; followed by S and then Se analogues
$-O-OH$	hydroperoxide
$>O$	ether or oxide
$>S$, $>SO$, $>SO_2$	sulfide, sulfoxide, sulfone
$>Se$, $>SeO$, $>SeO_2$	selenide, selenoxide, selenone
$-F$, $-Cl$, $-Br$, $-I$	fluoride, chloride, bromide, iodide
$-N_3$	azide

lists the commonest of these classes, again in priority order that
governs the choice if more than one of these classes of group is present
in one molecule. Examples are ethyl alcohol, ethyl chloride, phenyl
azide, and dimethyl sulfoxide.

Any substituents not denoted by the functional class name are
specified by the prefixes of substitutive nomenclature (*Table 4.3*) as,
for example, in *p*-bromobenzyl cyanide.

When a function is represented by a bivalent formula (e.g., $-O-$,
$>CO$), different groups attached at the two bonds are stated as separate
words in alphabetical order; or if the two groups are identical their
name is preceded by di-. Examples are benzyl phenyl ether, diethyl
ether, benzyl 1-naphthyl sulfide, and diethyl sulfoxide.

The reader may have noticed that '-carboxylic acid' or '-oic acid'*
is actually part of a radicofunctional name, since acid denotes the
function and the preceding word specifies the radical in adjectival form.
The reason why acids have to be included in substitutive nomenclature
is the extremely high priority for citation as suffix that is assigned to
them. Perhaps this is also the place to note that it is not the practice to
call organic acids by names such as hydrogen acetate which would
parallel the inorganic practice of hydrogen chloride, hydrogen
tetrachloroaurate, etc.

It is obvious that none of the *Tables 4.2* to *4.4* approaches completeness, for there are vague phrases such as 'their derivatives' and 'etc.'.
The variety in organic chemistry makes that inevitable for arbitrary
priorities such as these; completeness would require an alphabetical

---

* The difference between these names is explained in Chapter 6.

order or some form of computer logic. Yet it is surprising how very seldom the need for finer distinctions arises in practice.

Alphabetical Order

Reference has already been made several times to alphabetical order of prefixes in names, and it is time now to define it precisely. However, we must first distinguish two kinds of prefix. Those that are treated as an integral part of the parent compound name (e.g., cyclo-, iso-, oxa-, benzo-, etc.) are called non-detachable; those denoting substitution, which may be separated from the parent name and placed following it for indexing purposes, as in Benzene, chloro-, and 1-Butanol, 3-methyl-, are referred to as detachable. IUPAC rules permit hydro prefixes and those denoting subtraction or hetero bridge formation (e.g., epoxy) to be regarded as either detachable or non-detachable; but *CA*, which cannot tolerate alternative index names, normally treats epoxy and epithio as non-detachable but hydro prefixes and those denoting subtraction as detachable.

Detachable prefixes are arranged in alphabetical order, any multiplying parts of simple prefixes being neglected for this purpose. The atoms and groups are alphabetized first and the multiplying prefixes are then inserted, as in:

> *o*-bromochlorobenzene
> 4-ethyl-3-methyldecane
> 1,1,1-trichloro-3,3-dimethylpentane

A complex radical forms one prefix; it is therefore alphabetized under its first letter, as in:

> 1-(dimethylamino)-3-ethyl-4-(methylamino)-2-naphthoic acid
> 4-chloro-1,5-bis(dimethylamino)-3-ethyl-2-naphthoic acid

For otherwise identical prefixes the one with the lowest locant at the first cited point of difference is given first, as in:

> 1-(2-ethyl-3-methylpentyl)-8-(2-ethyl-4-methylpentyl)-
> naphthalene

As already noted (*see* p. 62), when two words serve similar purposes in a three-word name they are alphabetized, as in:

> ethyl methyl ketone
> butyl ethyl ether

Italics are neglected in alphabetizing, as in:

3-(*trans*-2-butenyl)-2-ethylphenol

Numbering

The principles of numbering (enumeration) used for compounds and for radicals are the same, with one exception, namely: for compounds the lowest available numbers are assigned to the principal (functional) group or groups; for radicals they are assigned to the 'free' valence or valences, and all substituent groups are then cited merely by prefixes.

'Lowest numbers' has a specific meaning in nomenclature. When two or more sets of numbers are compared in ascending order of numbers, that set is 'lowest' which contains the lowest individual number on the first occasion of difference*. Thus 1,1,7,8 is lower than 1,2,3,4 (*see also* examples **10** and **12**).

The IUPAC rules on the subject state that for aliphatic compounds lowest numbers are assigned successively, so far as applicable and until a final decision is reached, to:

(1)   principal groups,
(2)   unsaturation (i.e., double and triple bonds considered together),
(3)   double bonds,
(4)   triple bonds,
(5)   atoms or groups designated by prefixes,
(6)   prefixes in order of citation (in Great Britain and USA, alphabetical).

For ring systems there is a prescribed numbering (outlined in Chapter 5) which often leaves little or no room for choice according to the substituents present. However, so far as choice remains, lowest numbers are given to:

(1)   "indicated" hydrogen (*see* pp. 82, 117–119),
(2)   principal groups,
(3)   multiple bonds in compounds whose names indicate partial hydrogenation (cycloalkenes, pyrazolines, and the like),
(4)   prefixes,
(5)   prefixes in order of citation (which is alphabetical in USA and Great Britain).

*There has been a belief in some quarters that the total of numbers must be the smallest possible: that is erroneous.

The simplest case is when the numbering of the parent compound is entirely fixed in advance, as occurs with many cyclic compounds (*see* Chapter 5). Substituents in, say, quinoline, do not determine the numbering; compound **1** must be 6-chloro-3-methyl-2-quinolinecarboxylic acid.

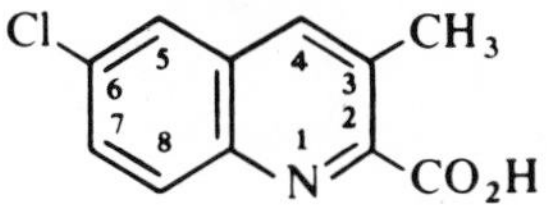

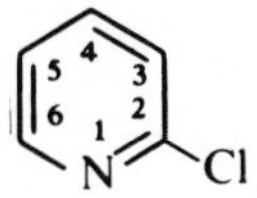

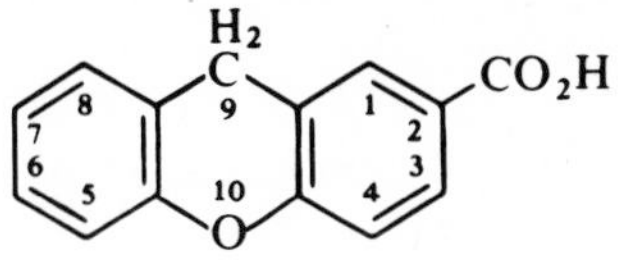

Next there is the simple case of one group (which need not be a principal group) in a compound such as **2, 3, 4**, or **5** which can be numbered in either of two ways: the lower numbers for substituents are as shown and are to be used.

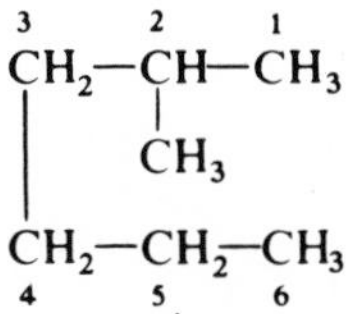

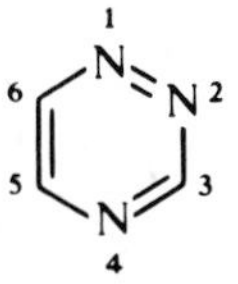

If the ring leaves the numbering completely free, e.g., in **6**, the principal group begins the numbering. If there is more than one principal group, they may be required together to decide the numbering, and may do so either wholly as in **7**, or partly as in **8**, where the methyl group decides which carboxyl group shall have number 1 and which number 3. Our definition of smallest numbers leads to decisions as shown in **9–13**.

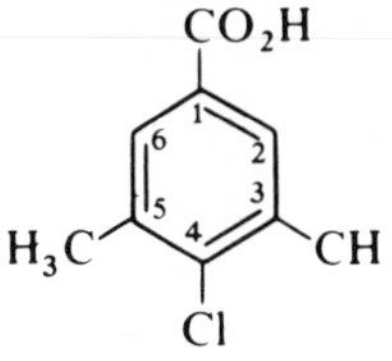

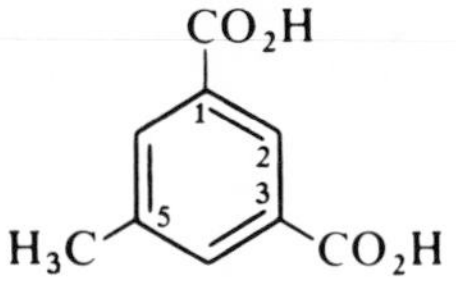

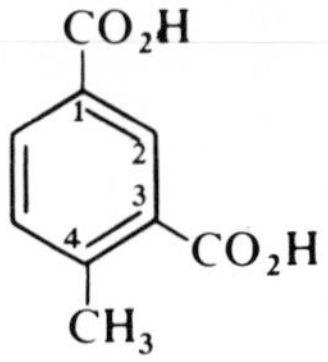

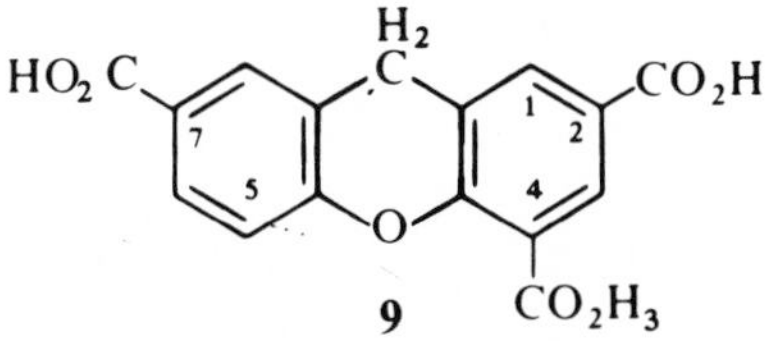

$CO_2H = 2,4,7-$ (not 2,5,7)

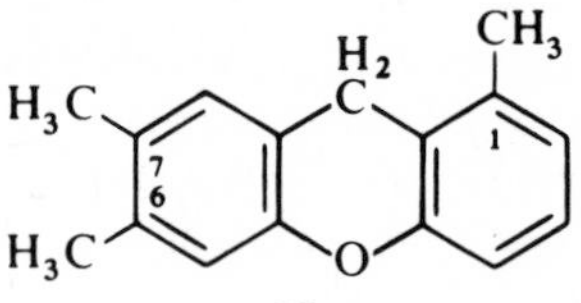

10

$CH_3 = 1,6,7$- (not 2,3,8)
(although $1 + 6 + 7 = 14$, and $2 + 3 + 8 = 13$ only*)

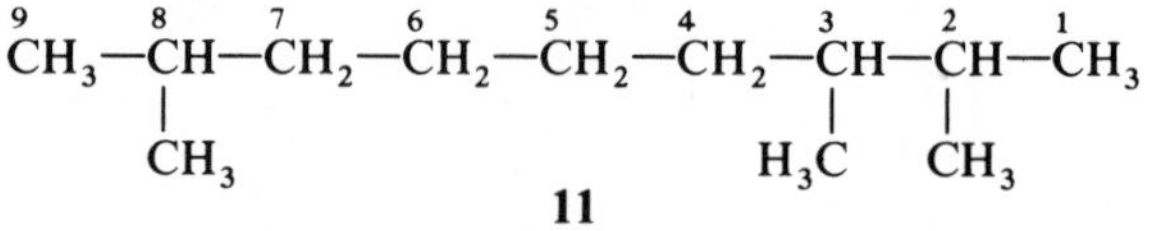

$$\overset{9}{C}H_3-\overset{8}{C}H-\overset{7}{C}H_2-\overset{6}{C}H_2-\overset{5}{C}H_2-\overset{4}{C}H_2-\overset{3}{C}H-\overset{2}{C}H-\overset{1}{C}H_3$$

with CH_3 below C-8, and H_3C, CH_3 below C-3, C-2

11

$CH_3 = 2,3,8$- (not 2,7,8-)

$$\overset{1}{C}H_3-\overset{2}{C}H-\overset{3}{C}H_2-\overset{4}{C}H_2-\overset{5}{C}H_2-\overset{6}{C}H_2-\overset{7}{C}H-\overset{8}{C}H-\overset{9}{C}H_2-\overset{10}{C}H_3$$

with CH_3 below C-2, and H_3C, CH_3 below C-7, C-8

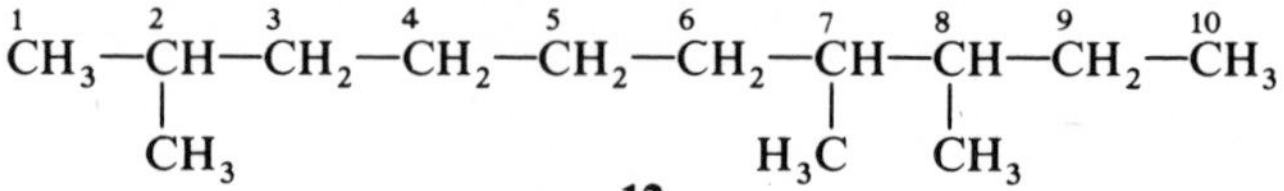

12

$CH_3 = 2,7,8$- (not 3,4,9-)
(although $2 + 7 + 8 = 17$, and $3 + 4 + 9 = 16$ only*)

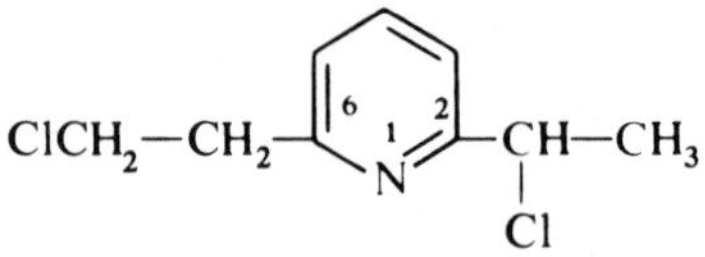

13

1-chloroethyl precedes 2-chloroethyl

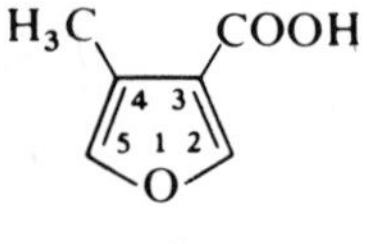

14

15

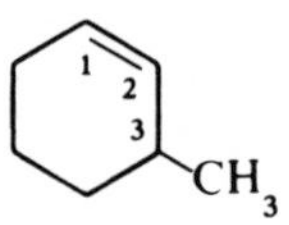

16

$$CH_3-CH_2-CH{=}CH_2$$

17

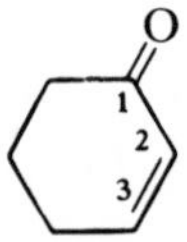

18

*As already noted, the total of numbers need not be the smallest possible.

Lastly we can run quickly through the IUPAC sequence. The principal group decides in cases such as **14** and **15**. If there is no principal group, a double bond cited as '-ene' decides, as in **16** and **17**; also it can make the second choice if the principal group is not decisive, as in **18**; note, however, that unsaturation not cited as -ene is not relevant here: 'hydro-' prefixes are treated in the same way as any substituent (*see* below). Certain heterocycles have modified names when partly hydrogenated (*see* pp. 94–96), and then too the unsaturation decides the numbering. If there is no principal group or '-ene' double bond, triple bonds are considered, as in **19** and **20**. Next, substituent prefixes are

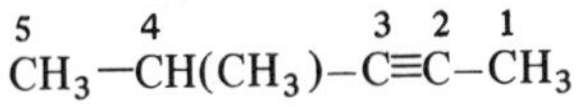

$$\overset{5}{CH_3}-\overset{4}{CH}(CH_3)-\overset{3}{C}\equiv\overset{2}{C}-\overset{1}{CH_3}$$

19

$$\left(\overset{2}{C}\equiv\overset{1}{C}\atop [CH_2]\right)_x$$

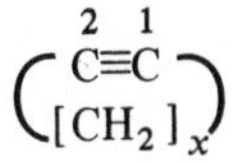

20

considered, first all together, independently of the kind and including hydro-. Examples are **21** (1,2,4,5,8-, not 1,4,5,6,8-), **22** (1,2,3,4-tetrahydro-), and **23** (1,2,5,8-, not 1,4,5,6-). All other things being equal, the lower number goes to the prefix cited first in the name, as in **24** (bromo before chloro), **25** (ethyl before methyl), **26** (dimethylamino before methyl), and **27** (tetrahydro before tetramethyl).

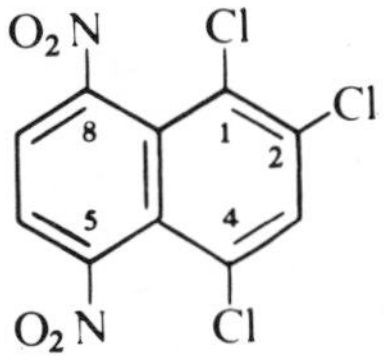

21

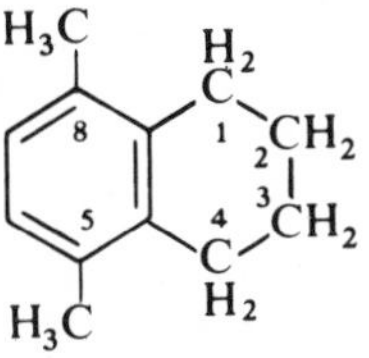

22

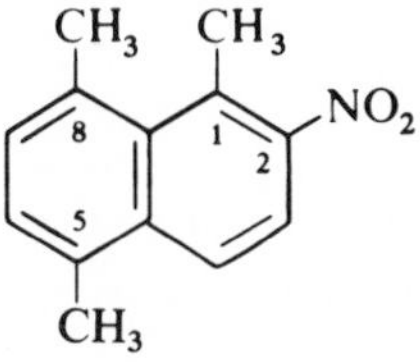

23

24

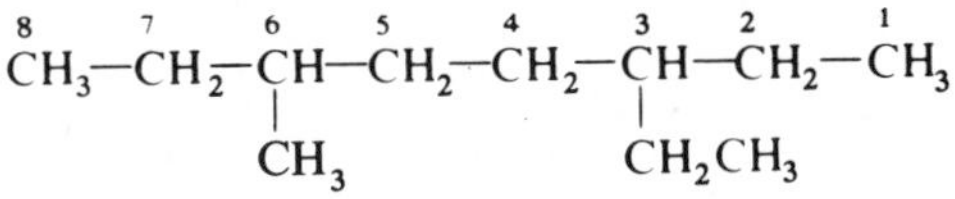

$$\overset{8}{CH_3}-\overset{7}{CH_2}-\overset{6}{CH}-\overset{5}{CH_2}-\overset{4}{CH_2}-\overset{3}{CH}-\overset{2}{CH_2}-\overset{1}{CH_3}$$

with CH_3 on C6 and CH_2CH_3 on C3

25

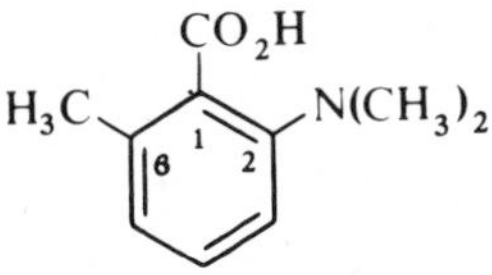

26

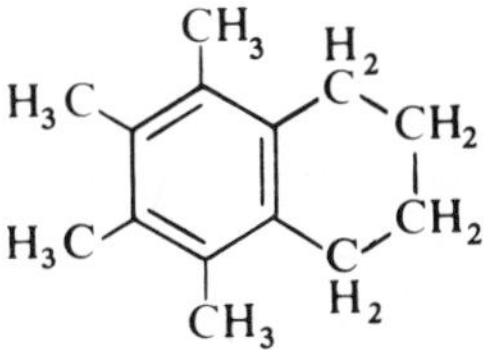

27

That completes the general rules. Problems, largely theoretical, arising from severe ramification or multiple unsaturation of aliphatic chains are considered in the IUPAC rules[1b] but need not concern us here.

Conjunctive Nomenclature

This type of nomenclature (earlier called 'additive nomenclature' in *CA*) can be applied when a ring system is attached through carbon or nitrogen to a carbon atom of an aliphatic chain that bears a principal group. The name of the ring system is followed directly (without space) by the name of the aliphatic chain and principal group, as in 2-naphthaleneethanol. Two hydrogen atoms are assumed to be lost in the linking process, and not one as in substitutive nomenclature. Locants used for the ring system are the usual numerals – the '2' in the example above refers to the 2-position of the naphthalene nucleus and not that of the ethanol. For this system of nomenclature the aliphatic chain runs from the ring system to the functional group, but not beyond, at either end; all groups attached to this chain are treated as substituents, and Greek letters are used as locants for them, starting with α for the carbon atom bearing the principal group. Substituents, whether on the aliphatic chain or the cyclic component, are cited as prefixes. Thus, for example, we arrive at 2-naphthalenepropionic acid **28**, 2-naphthaleneethanol **29**, α,γ-dimethyl-3-pyridinepropanol **30**, and 1,2,3-cyclohexanetriacetic acid **31**.

$$2\text{-}C_{10}H_7\text{—}\overset{\beta}{C}H_2\overset{\alpha}{C}H_2CO_2H \qquad 2\text{-}C_{10}H_7\text{—}\overset{\beta}{C}H_2\overset{\alpha}{C}H_2OH$$

28

29

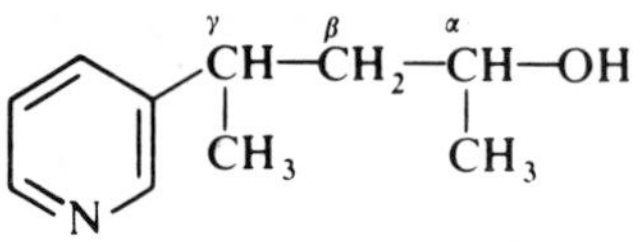

30

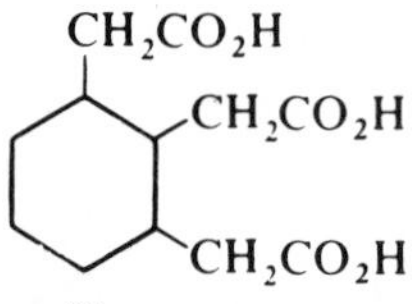

31

If the side chains are not identical, one (the simpler) is named as prefix. For example, **32** would be called 3-(2-carboxyethyl)-2-naphthalenebutyric acid.

This method is subject to certain limitations, which are defined differently by IUPAC and *CA*. Thus the IUPAC rules do not permit application to a benzene ring that is only monosubstituted; *CA* does not use the method when the ring system and the aliphatic chain are linked by a double bond, as in **33** (IUPAC name: indene-$\Delta^{1,a}$-acetic acid, wherein the Δ cites the double bond between the ring position 1 in indene and the a position in acetic acid). Both restrict application when the aliphatic chain is unsaturated or difunctional. IUPAC (only)

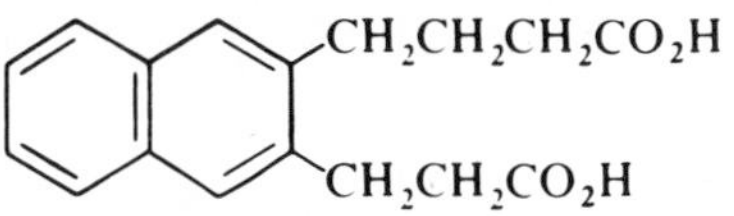

32 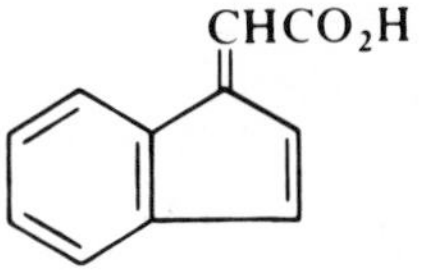   **33**

authorizes the use of $-NHCO_2H$ as the aliphatic chain. Further detail on rare cases will be found in the 1970 IUPAC rules[1c].

A Closer Approach to a Name

The ways in which a name should be built were given in outline on pp. 57–59). Now that principal groups and numbering have been discussed, the outline can be filled in.

The starting point remains the principal group, as it was for numbering. This identity is very important, for it ensures that the name and the numbering shall run on parallel lines. (If there is no principal group in the compound, the name is approached simply as described in the next chapter, and prefixes are added later.)

For aliphatic compounds the complete chain is the parent to which the suffix denoting the principal group is added, as in 2-hexanol. If there is a choice of chains, priority is given by IUPAC to the most unsaturated as in 2-propyl-2-buten-1-ol **34**, even though this may not be the longest. The *CA* name, 2-ethylidene-1-pentanol, does not present this situation.

$$CH_3CH{=}CCH_2OH$$
$$|$$
$$CH_2CH_2CH_3$$

34

$$CH_3CH_2CHCH_2OH$$
$$|$$
$$CH_2CH_2CH_3$$

35

If there is a choice between chains of equal degree of hydrogenation, then, of course, the longest chain is chosen, as in 2-ethyl-1-pentanol **35**, even though as a result the related compounds **34** and **35** are then named from different parent hydrocarbons.

It will be remembered that the principal group is the group highest in *Table 4.3* (p. 61), and that all others, including 'hydro-'*, are given as prefixes, all in alphabetical order. This division into suffix and prefixes can cause chemically unwelcome changes in name or numbering, or both: for instance, compound **36** is 3-hexanol, but **37** is 4-hydroxy-3-hexanone. The same sort of change, can, however, result from shift of a double bond: **38** is 2-ethyl-1-hexene (but 3-methyleneheptane in *CA*), but **39** is 5-methyl-2-heptene. The fact is that names are founded on structures, not on chemical relations between two substances; the reason, of course, is that it is often easy to find relationship to more than one other substance so that choice of a name would then become ambiguous.

$$\underset{OH}{\overset{3\ 2\ 1}{CH_3\,CH_2\,CH_2\,\underset{|}{C}HCH_2\,CH_3}}$$

36

$$\underset{O\ \ OH}{\overset{1\ \ 2\ \ 3\ \ 4}{CH_3\,CH_2\,\underset{\|}{C}-\underset{|}{C}HCH_2\,CH_3}}$$

37

$$\underset{CH_2\,CH_3}{\overset{3\ \ 2\ \ 1}{CH_3\,CH_2\,CH_2\,\underset{|}{C}H_2\,\underset{|}{C}=CH_2}}$$

38

$$\underset{CH_2\,CH_3}{\overset{1\ \ \ 2\ \ \ 3\ \ 4\ \ 5}{CH_3\,CH=CHCH_2\,\underset{|}{C}HCH_3}}$$

39

There is no difficulty in citing multiple groups of one kind attached to a single chain or ring system; 3,4-hexanediol and 1,2,3,4-benzene-tetracarboxylic acid admit no argument. With branched chains further rules come into play. Substance **40** must be named so that both functional groups can be cited as suffixes to the name of the main chain: the name is 2-butyl-1,4-pentanediol and not 2-(2-hydroxypropyl)-1-hexanol, even though the latter involves a hexane chain and the correct name involves only a pentane chain.

$$\underset{CH_2\,CH(OH)CH_3}{CH_3\,CH_2\,CH_2\,CH_2\,\underset{|}{C}HCH_2\,OH}$$

40

$$\underset{CH_2\,CH_2\,OH}{HOCH_2\,CH_2\,\underset{|}{C}HCH_2\,CH_2\,OH}$$

41

The principle of the preceding paragraph is often spoken of as 'treating like things alike'. In general that is indeed a good thing to do; but it cannot be made into a guiding light for nomenclature, for there

---

*This does not apply in German.

is often argument about degrees of 'likeness', and in any case the principle itself must often be discarded. Consider, for example, the simple triol **41**. Here the name must be based on a pentane chain containing two hydroxyl groups, leaving the third hydroxyl group in a side chain: the correct name is 3-(2-hydroxyethyl)-1,5-pentanediol. Now there could be a system of nomenclature by which this substance would be called 2-ethylpentane-1,5,2'-triol, but in fact such a system is not used and this name is incorrect.

Wholly symmetrical aliphatic compounds such as **41** are, however, rather rare. More commonly, branched chains offer more alternatives. Consider a choice between various chains containing the same number of identical groups. As when there was only one principal group, the additional rule favouring unsaturation requires that compound **42** be named 4-(4-hydroxybutyl)-2-octene-1,8-diol, even though selection of the two saturated segments to constitute the main chain would give the

$$HOCH_2CH_2CH_2CH_2\underset{\displaystyle \overset{|}{CH_2CH_2CH_2CH_2OH}}{CH}CH=CHCH_2OH$$

42

name 5-(3-hydroxy-1-propenyl)-1,9-nonanediol which *CA* uses. When there is no difference in degree of saturation the longer chain would be chosen, and the saturated analogue of **42** would be called 5-(3-hydroxypropyl)-1,9-nonanediol.

There are occasions when further choices are needed for highly branched chains, but they become too rare to need discussion here. However, before passing on, let us note that the acid **43** can be called butylsuccinic acid or 2-butyl-1,4-butanedioic acid; whether the trivial succinic or the systematic butanedioic is used, the same principles are applied, leading here to the C_5 chain, not the C_7 chain, as the basis of the name.

$$CH_3CH_2CH_2CH_2\underset{\displaystyle \overset{|}{CH_2CO_2H}}{CH}CO_2H$$

43

44

The rule about seniority for principal groups, in the order on p. 60, is independent of the relative numbers of the various groups. The compound **44** is 3,4-dihydroxycyclohexanone, and not 4-oxo-1,2-cyclohexanediol: this is in line with the numbering rules which give the lowest available number to the principal group even though the sequence 1,2,4 is lower than 1,3,4. There are many and varied types of problem in which this consideration arises: but the answer is always the same, so further elaboration is unnecessary.

The procedures discussed above are applicable equally when the principal groups are attached directly to cyclic systems. With the more complex ring systems, derivation of the full name is, in fact, usually simpler (once the cyclic parent has been named) because the numbering is at least partly fixed. With simpler ring systems, and particularly the aromatic ones, the systematic procedures become more troublesome, partly from older custom and partly because there is here a very large number of trivial names which should reasonably be carried over to the derivatives.

Strict application of the rule that principal groups shall be cited as suffix is relatively recent, particularly for cyclic compounds, and there are many situations where the resulting names may appear unfamiliar to, or even shock, some older chemists, 2,3-Dihydroxynaphthalene and 2-aminophenanthrene, for instance, may appear more usual to some than the correct 2,3-naphthalenediol and 2-phenanthrenamine. And there is a further stumbling block with derivatives of N-heterocyclic systems; the '-ine' of, say, pyridine may be considered a suffix prescribing the principal group, but if a 'senior' group in the sense of *Table 4.3*, say $-OH$ or $-CO_2H$, is present, then that should be stated as suffix, a procedure that gives in these cases, e.g., 2-pyridinol and 4-pyridine-carboxylic acid. These are IUPAC and *CA* names, so it seems wrong to regard the -ine of pyridine as a suffix. If that is so, then names such as 2-pyridinamine become perfectly satisfactory, and that is indeed the case; that is the modern name whose use is advisable.

With compounds containing both chains and rings the matter is simple when the functional groups are directly attached either all to the chain or all to the ring system; these groups decide the parent name, e.g., m-propylphenol but 3-phenyl-1-propanol (*CA*, benzenepropanol). As this shows, conjunctive nomenclature should never be forgotten. Difficult cases can arise when there are principal groups attached to both the chain and to the ring system, as they cannot all be treated as suffixes: the ring system or the chain must be chosen as parent, and the choice will usually fall on that part that is the more complex or the more important chemically, even though these criteria are both matters of opinion. Few would quarrel with names such as 8-(p-hydroxyphenyl)-2,4,6-octatrien-1-ol or 1-O-(p-hydroxyphenyl)glucose; but it may depend on the chemistry under discussion whether a name such as 3-(p-hydroxyphenyl)-1-propanol, 4-hydroxybenzenepropanol (*CA*) or p-(3-hydroxypropyl)phenol is chosen.

Care is similarly required when principal groups are present in more than one chain or in more than one ring system forming part of a single compound. IUPAC has laid down criteria as to the 'seniority' between various chains and between various rings; in general the more complex is the senior, but the rules[1d] must be studied for detail.

Finally there is the nomenclature that takes advantage of the fact that organic synthesis often leads to symmetrical compounds of type X–Y–X where the same principal group is present in both X groups but not in Y. When X and Y are not both aliphatic or both cyclic, or when Y is a hetero group, simple and readily intelligible names can be devised by a special elaboration of the standard rules, as mentioned briefly on p. 58. The examples given were **45** and **46**, in **45** the two

$$p\text{-}HO_2CCH_2-C_6H_4-CH_2CO_2H \ and \qquad\qquad \textbf{45}$$

$$H_2NCH_2CH_2-O-CH_2CH_2NH_2 \qquad\qquad \textbf{46}$$

acetic acid residues are cited as acetic acid and 'doubled' by the prefix 'di-'; the C_6H_4 group called phenylene is then named first of all, giving p-phenylenediacetic acid, which describes the structure as clearly as phenylacetic acid describes $C_6H_5CH_2CO_2H$. Similarly, compound **46** is called 2,2'-oxydi(ethylamine), where the parentheses differentiate the two ethylamine residues from diethylamine $(C_2H_5)_2NH$. In this kind of nomenclature *CA* now uses bis-, tris-, etc., instead of di-, tri-, etc., and encloses the name following that infix in parentheses whether it is complex or not: p-phenylenebis(acetic acid), 2,2'-oxybis(ethanamine). This procedure is quite widely applicable, for a variety of groups named as described on pp. 55, 58 can function as Y: for example, $>$CO carbonyl, $-CH_2-$ methylene, $-CH_2CH_2-$ ethylene, $-CH_2CH_2CH_2-$ trimethylene (etc.), $-NHCONH-$ ureylene, $-S-$ thio, $>$CS thiocarbonyl, $>$NH imino, $-N=N-$ azo. Further, the principle can be extended to tervalent radicals, e.g., $N(CH_2CO_2H)_3$ is nitrilotriacetic acid; finally, the Y group may itself be complex and symmetrical as in **47**, which is named 4,4'-methylenedioxydibenzoic acid. A warning is, however,

$$HO_2C-\!\!\!\!\bigcirc\!\!\!\!-O-CH_2-O-\!\!\!\!\bigcirc\!\!\!\!-CO_2H$$

47

needed: trouble will arise if there is not complete symmetry of the two X groups or within Y, and this nomenclature should not be attempted in such cases.

It will be clear, then, how the building of a name is not always simply a following of rules. Though in a truly surprisingly high proportion of cases the rules and common trivial or simple systematic names suffice, and though, let it be emphasized again, rules should not be broken without good cause, yet chemistry must not be forgotten. The versatility of organic nomenclature in fact arises from the reasonable wish to express chemical relations in a name; and in this too the difficulties originate, as when multiple chemical relations lead to conflicting types of name or when a specialized usage is extended beyond its appropriate field. Then it is wise to revert to the systematic types, which thus remain the backbone of nomenclature. It is essential to appreciate both the uses and the limitations of systematics before indulging in excursions from it.

REFERENCES

1. *IUPAC Nomenclature of Organic Chemistry, Definitive Rules for: Section A. Hydrocarbons; Section B. Fundamental Heterocyclic Systems; Section C. Characteristic Groups Containing Carbon, Hydrogen, Oxygen, Nitrogen, Halogen, Sulfur, Selenium and/or Tellurium*, 1969. A, B 3rd Ed.; C 2nd Ed.; Butterworths, London (1971); now available from Pergamon Press, Oxford. (*a*) pp. 123–127; (*b*) pp. 8–11, 13–15; (*c*) pp. 119–123; (*d*) pp. 97–105
2. *Chemical Abstracts, Index Guide to Vol.* 76, p. 291 (1972); *Ninth Collective Index, Vols.* 76–85 (1972–1976), *Index Guide*, para. 127

5

Organic: Hydrocarbons and Heterocycles

Acyclic Hydrocarbons

Standard nomenclature of alkanes is well known, but a few special points are worth noting.

The prefix 'iso', to denote terminal $(CH_3)_2CH-$, is restricted to C_4-C_6 parent alkanes and '*tert-*' to C_4 and C_5 alkanes; '*sec-*' is acceptable in *sec*-butyl but not at all for hydrocarbons, and 'neo' only in neopentane, $C(CH_3)_4$. It should be noted that '*tert-*' and '*sec-*' are italicized and followed by a hyphen, whereas 'iso' and 'neo' are not; '*n-*', used to specify *normal* (which should not be done except for contrast), should be italicized and followed by a hyphen as it is not to be used in indexing. These are IUPAC rules; *CA* uses no such prefixes in naming alkanes and, for example, indexes $C(CH_3)_4$ at Propane,2,2-dimethyl-.

In highly branched alkanes choice of the main chain(s), i.e., that chain which is regarded as substituted by the other chain(s), can present considerable difficulties, at least in hypothetical compounds. Complex rules have been provided by IUPAC to cover such cases[1a].

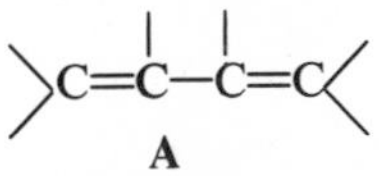

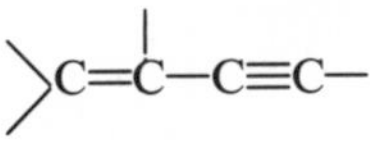

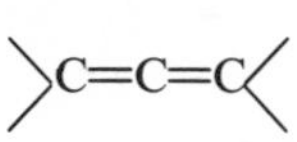

Unsaturation, as is well known, is indicated by replacing the ending '-ane' by '-ene' (for C=C) or '-yne' (for C≡C), and both may occur in a name ('-ene' before '-yne', with elision of '-e' before a vowel), e.g., 1-buten-3-yne or 2-hexadecene-5,7-diyn-4-ol.

Compounds containing groups **A** (or more extended forms) are said to have conjugated double bonds; **B** groups are also said to have conjugated multiple bonds. Compounds containing the group **C** or similar but more extended systems are said to have cumulative unsaturation and are generically called cumulenes; they are numbered normally, e.g., 1,2-butadiene $CH_3 CH=C=CH_2$.

Cycloalkanes

The nomenclature of cycloalkanes and their unsaturated derivatives follows closely that of unbranched alkanes; all substituents attached to the ring are named as such. Their numbering follows the principles discussed in Chapter 4.

Benzene Derivatives

The IUPAC rules follow general custom; they authorize the use of seven trivial names as parents as shown in *Table 5.1. Chemical Abstracts*

Table 5.1 BENZENE AND ITS SIMPLE HOMOLOGUES

Benzene   Cumene 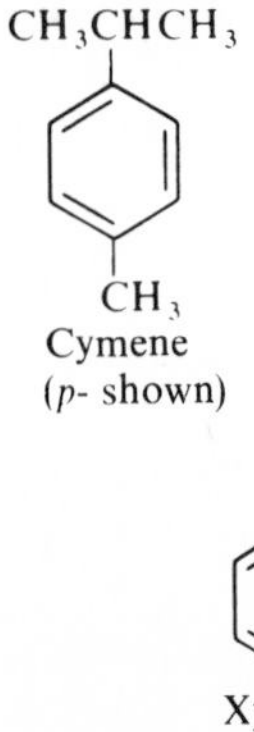   Cymene (*p*- shown)   Mesitylene

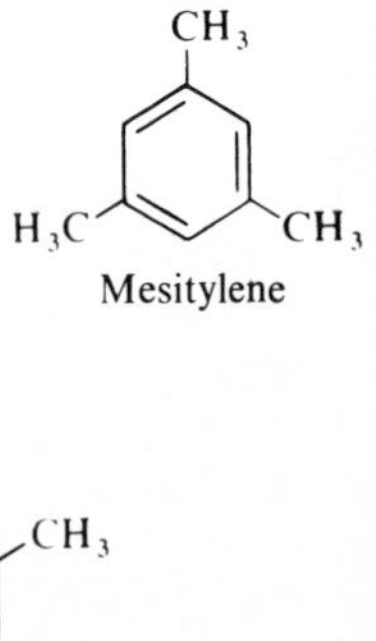

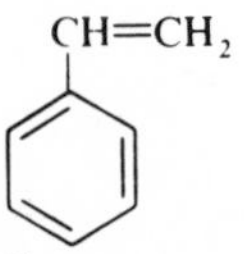

Styrene 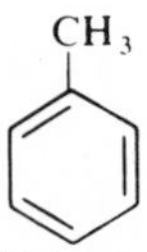   Toluene 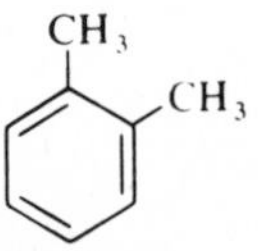   Xylene (*o*- shown)

uses only benzene, and names the others as substituted benzenes. The generic name for aromatic hydrocarbons is arene.

The highly reactive unisolated intermediate **1** is now generally called benzyne, and its derivatives and analogues are named accordingly.

1

Bi-Compounds

Bi- has long been American usage to denote direct union of two identical rings, as in **2** and **3**, which in Europe were earlier called diphenyl and 2,2′-dinaphthyl. The IUPAC rules early adopted 'bi-' (in place of di-), and this prefix is now generally used; it is exclusive and unique for 'doubled groups' (except for bicyclo, *see* below). The IUPAC rules, however, permit doubling of either the group, giving binaphthyl, or of the molecule, giving binaphthalene; in the latter case the use of 'bi-' carries the implication that two hydrogen atoms are lost as in conjunctive nomenclature (*see* p. 68). Doubling of the molecule is now almost universally preferred; but so is biphenyl, as an exception of very long standing (note that the radical $C_6H_5C_6H_4-$ is biphenylyl, the second '-yl' being the termination for a radical).

2   **3**

$$HO_2C-\text{⟨4 1⟩}-\text{⟨1′ 4′⟩}-CO_2H$$

4

Numbering is as shown in formulas **2** and **3**, the points of union having the lowest available numbers.

The 'bi' nomenclature is reserved for unsubstituted compounds in the sense that only parent compounds may be 'doubled'; for example, the IUPAC name for compound **4** is 4,4′-biphenyldicarboxylic acid, not *p,p′*-bibenzoic acid (as it was earlier in America).

Ter, quater, etc., are used for larger ring assemblies, always with the name of the molecule (e.g., ternaphthalene, quaterpyridine), but again with terphenyl, quaterphenyl, etc., as exceptions. For assemblies of the same unit, unprimed numbers are assigned to one of the terminal units, the nearest neighbour unit has singly primed and the next doubly primed numbers, and so on. As usual, points of union have the lowest numbers possible, these numbers being considered an attribute of the parent structure and not subject to change by presence of a suffix — e.g., the CO_2H groups of **4** are numbered 4 and not 1.

Polycyclic Hydrocarbons

The nomenclature of polycyclic hydrocarbons, which is the basis also for that of heterocycles, is complex because of the multitude of compounds and different situations that can be met. Chemists should become familiar with the basic principles, which are all that can be described here. There is a considerable list of prescribed names, a method (the fusion method) of building names for more complicated systems derived from members of that list, a second method of building polycyclic names (the bicyclo system) independent of the fusion principle and the prescribed list, plus a few special methods for specific types of compound. We must deal with them in that order, but, before doing that, it will be best to note the standard method of orienting and numbering polycyclic compounds:

(1)   Whenever possible, rings are drawn with two sides vertical (a three-membered ring with one vertical).
(2)   As many rings as possible are then drawn in a horizontal line, irrespective of their size.
(3)   As much as possible of the remainder of the formula is arranged in the top right quadrant, and as little as possible in the bottom left quadrant (the middle of the first row being the centre of the circle).
(4)   *Then*, numbering proceeds clockwise round the periphery, starting in the right-hand ring of the top row at the first carbon atom not engaged in ring fusion. Atoms engaged in ring fusion (i.e., at the 'valley' positions) receive roman letters a, b, etc., after the numerals of the preceding atoms. (Subsidiary rules legislate for less common points.)
    Examples are chrysene **5** and rubicene **6**.

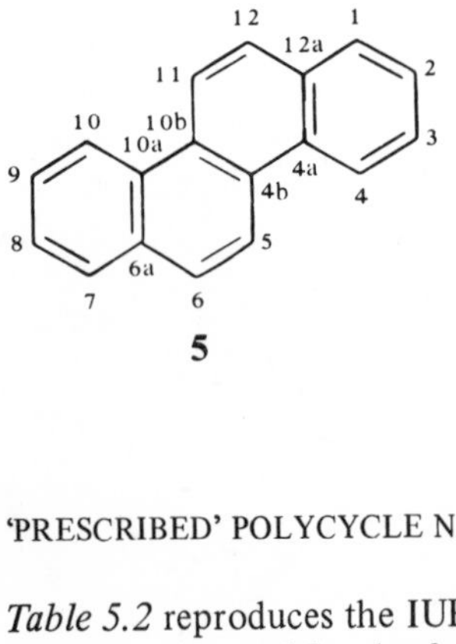

5

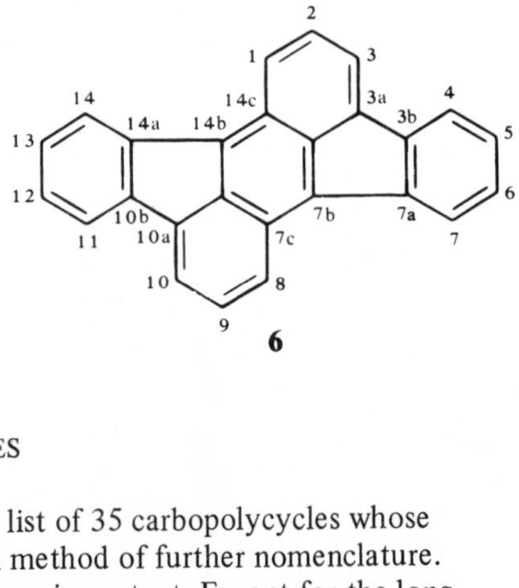

6

'PRESCRIBED' POLYCYCLE NAMES

Table 5.2 reproduces the IUPAC list of 35 carbopolycycles whose names can be used for the fusion method of further nomenclature. A few points concerning this list are important. Except for the long-established anthracene and phenanthrene, the numbering is that derived by the standard procedure just given. All the compounds are in the oxidation state where they contain the maximum number of noncumulative double bonds, and the ending '-ene' of the names of these ring systems

79

Table 5.2 HYDROCARBONS WHOSE NAMES ARE PRESCRIBED BY
IUPAC AS AVAILABLE (IN ORDER OF INCREASING 'SENIORITY'*) FOR
RING FUSION, NUMBERING IS IN ACCORD WITH IUPAC AND THE
RING INDEX

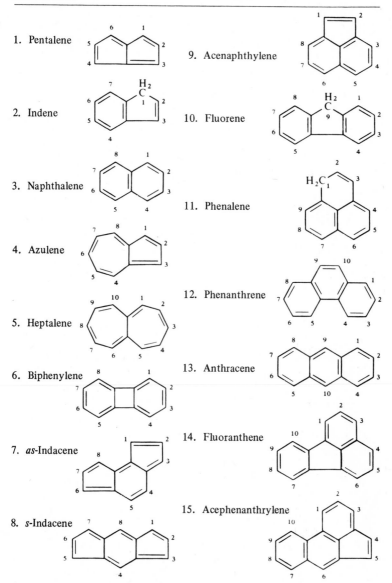

1. Pentalene
2. Indene
3. Naphthalene
4. Azulene
5. Heptalene
6. Biphenylene
7. *as*-Indacene
8. *s*-Indacene
9. Acenaphthylene
10. Fluorene
11. Phenalene
12. Phenanthrene
13. Anthracene
14. Fluoranthene
15. Acephenanthrylene

*By 'seniority' is here meant preferred choice (if any) for citation as parent
(i.e., last-named component in the fusion name).

continued

80

Table 5.2 (*continued*)

16. Aceanthrylene

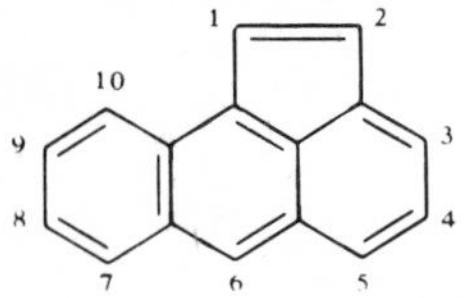

17. Triphenylene

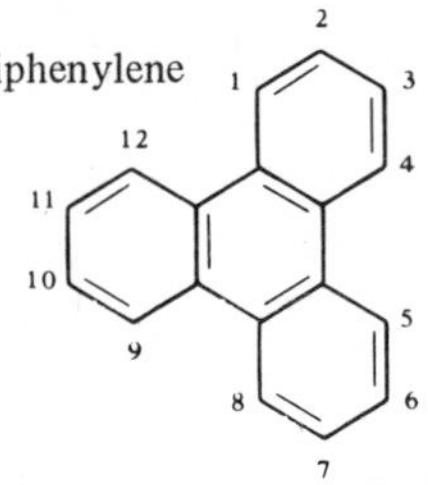

18. Pyrene

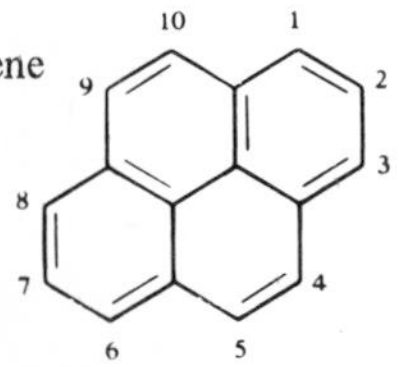

19. Chrysene

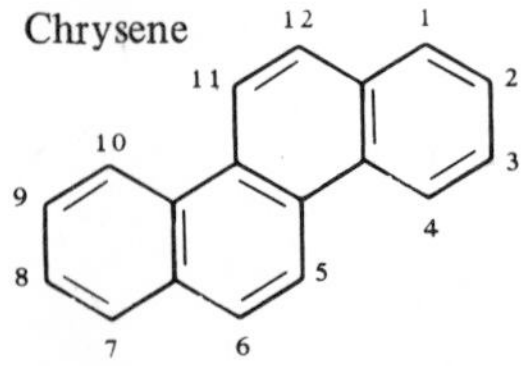

20. Naphthacene

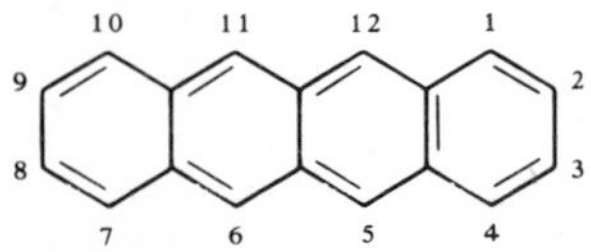

21. Pleiadene

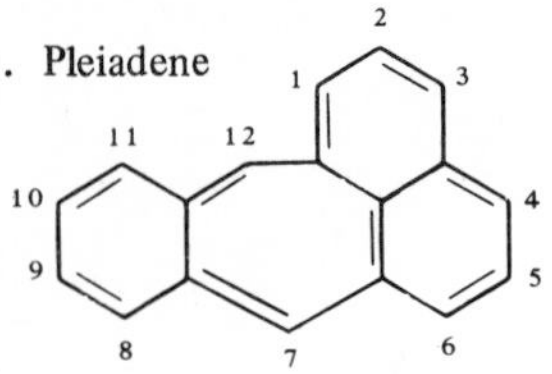

22. Picene

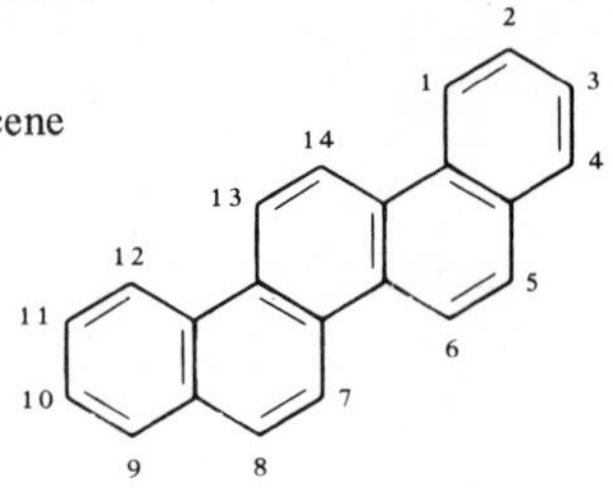

23. Perylene

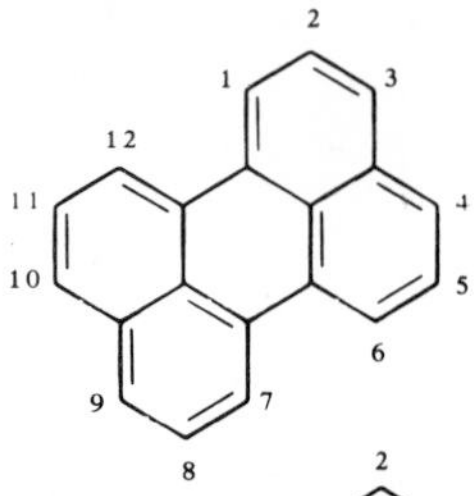

24. Pentaphene

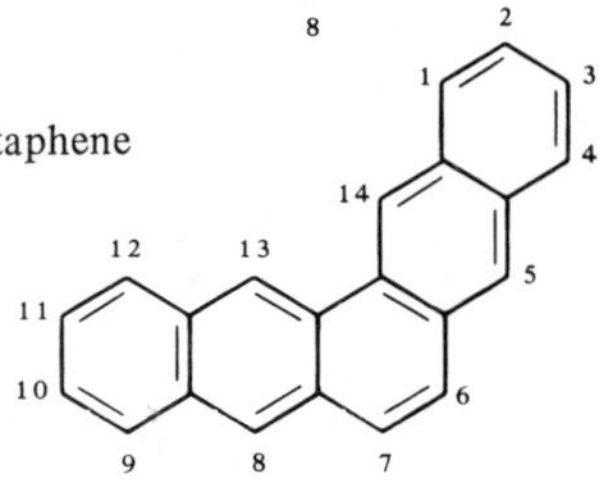

25. Pentacene

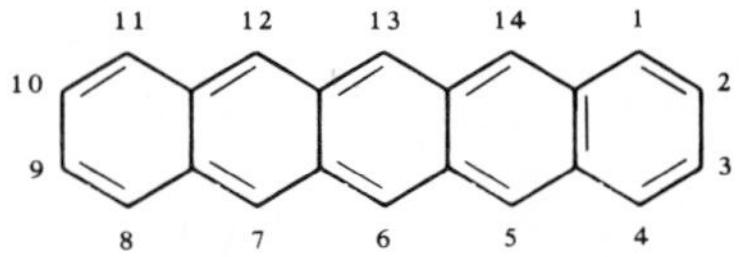

26. Tetraphenylene

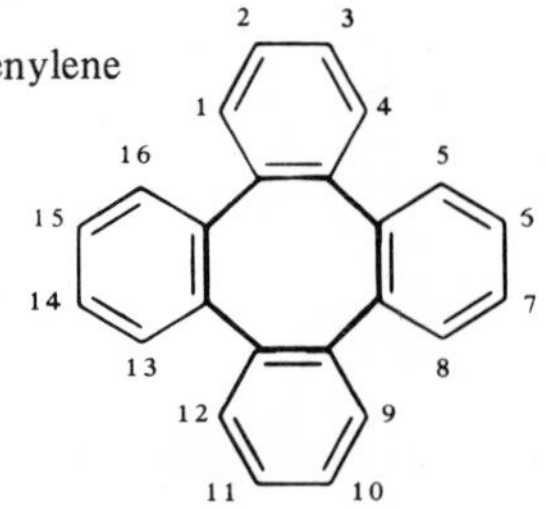

continued

Table 5.2 (*continued*)

27. Hexaphene

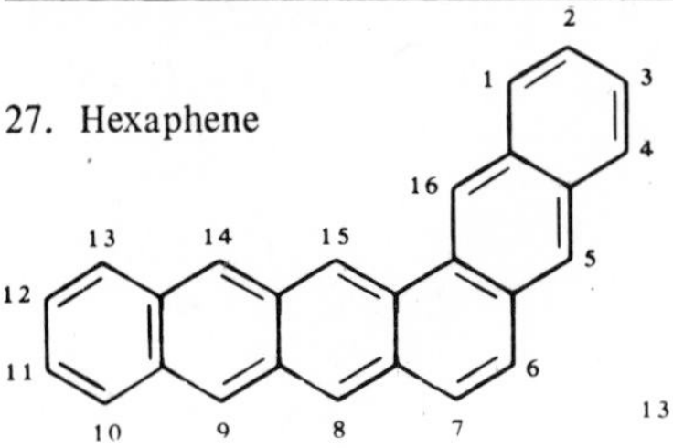

28. Hexacene

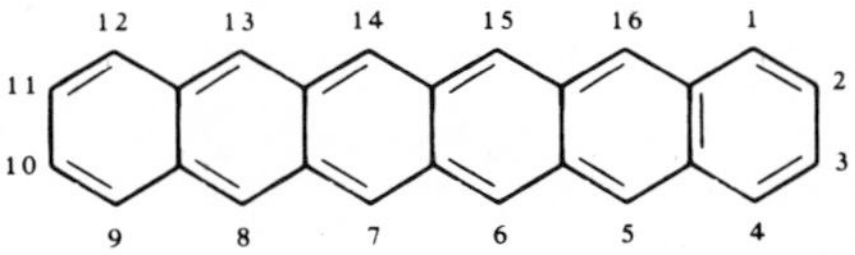

29. Rubicene

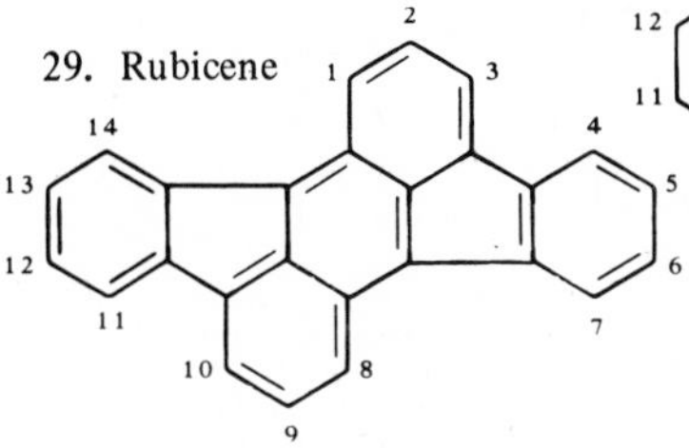

30. Coronene

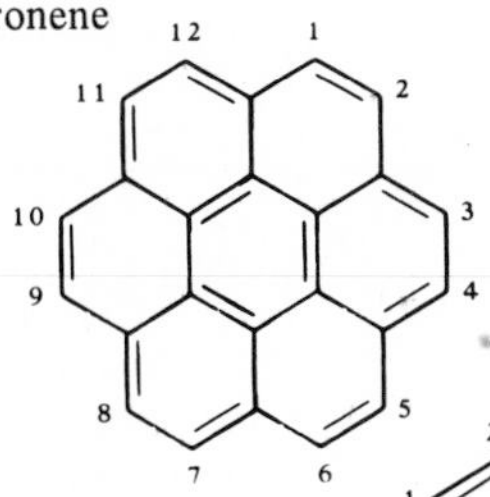

31. Trinaphthylene

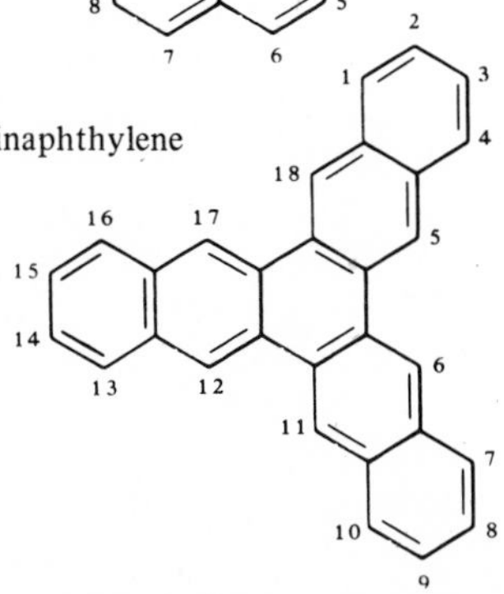

32. Heptaphene

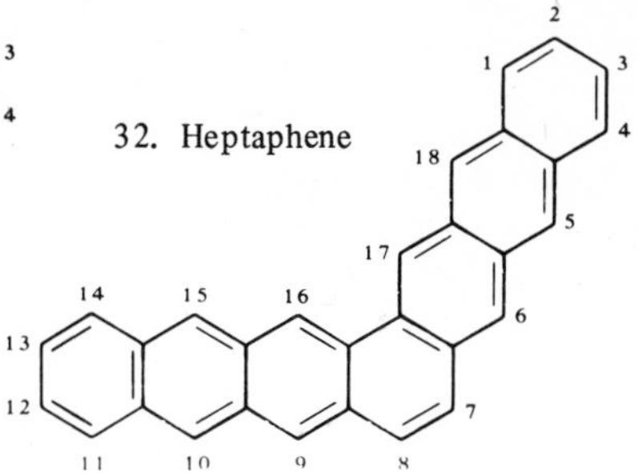

33. Heptacene

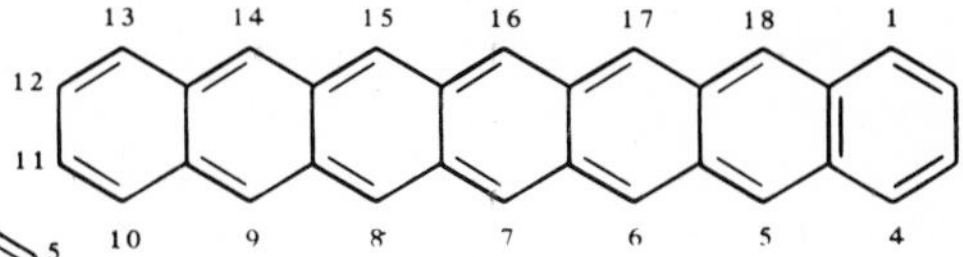

34. Pyranthrene

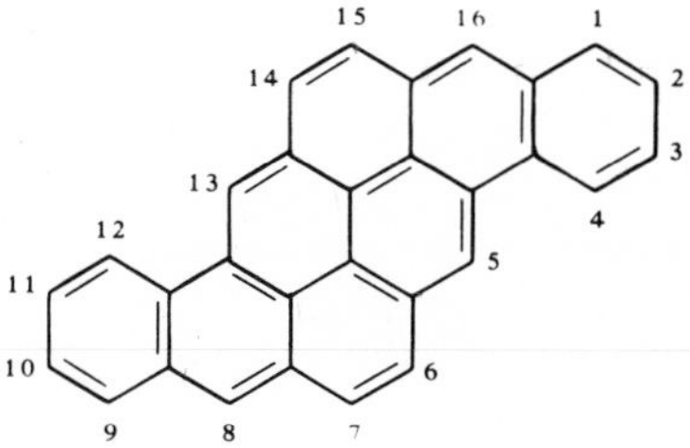

35. Ovalene

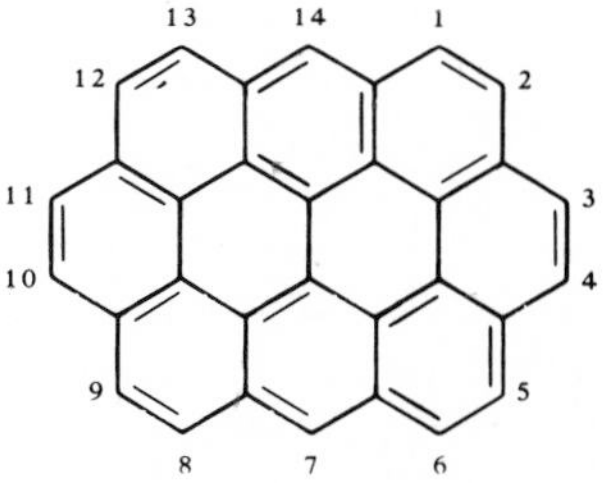

should be understood to denote this (and not just one double bond as for aliphatic compounds). Some cyclic skeletons require a CH_2 component in the ring; these components are to be indicated by italic *H* preceded by a numeral, except when in common compounds such as indene and fluorene its position can be assumed to be the normal one (*see Table 5.2*). If the CH_2 group were in a different position, say, as in compound 7, the name would be 3*H*-fluorene. Such hydrogen is termed

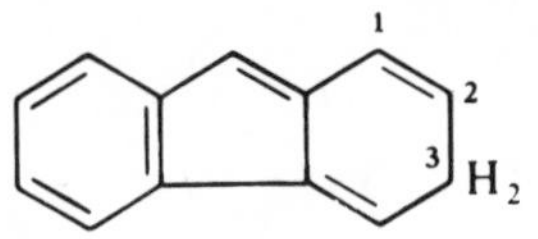

7

'indicated hydrogen'; it is discussed in greater detail on pp. 117, 125. *Chemical Abstracts* now always specifies indicated hydrogen even for the commonest form, and names compound 2 of *Table 5.2* as 1*H*-indene, compound 10 as 9*H*-fluorene, and compound 11 as 1*H*-phenalene.

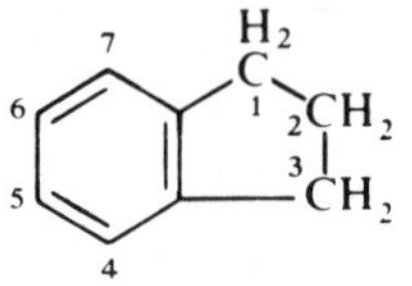

8 Indan 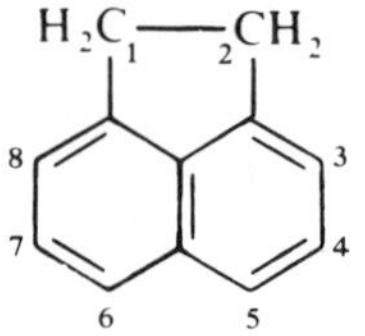   **9** Acenaphthene 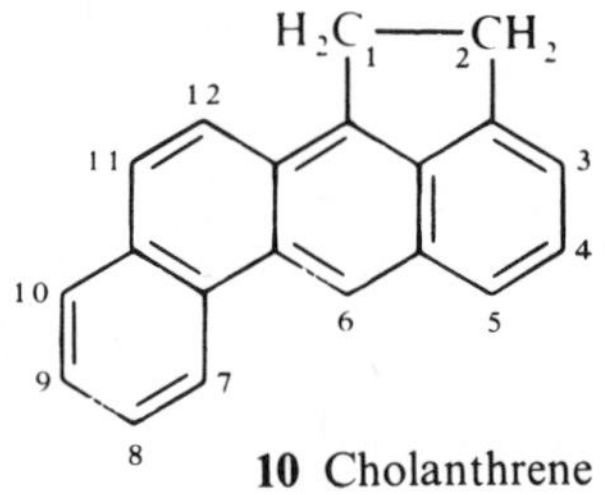   **10** Cholanthrene

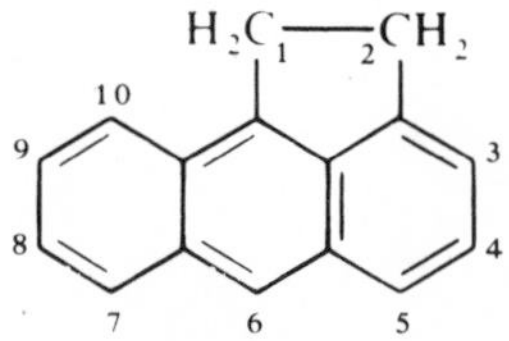

11 Aceanthrene 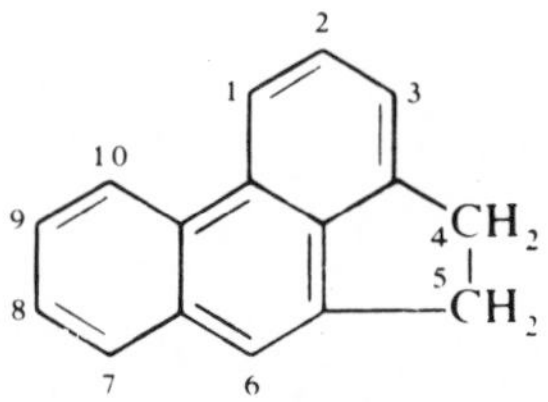   **12** Acephenanthrene

Hydrogenation is normally denoted by 'hydro-' prefixes (including 'perhydro-' for complete hydrogenation), but the seven partly hydrogenated systems 8 to 14 have IUPAC-recognized specific names (not to be used for ring fusion, and not used by *CA* indexes). Note the relation of indan to indene and of the ace- . . . -ene to the ace- . . . -ylene names.

Table 5.2 includes six compounds whose names have been systematized starting from the trivial name naphthacene (four benzene rings

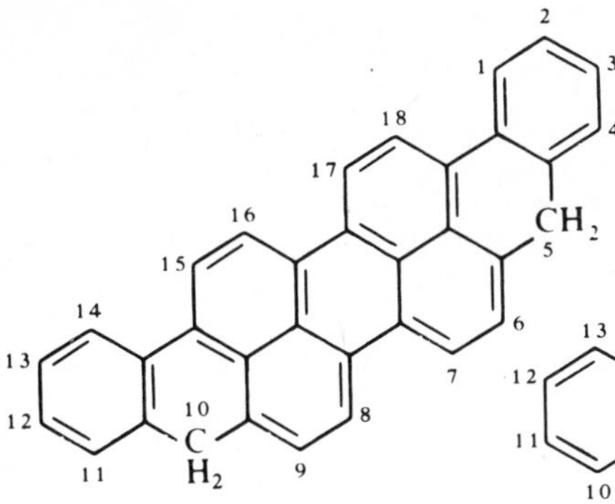

13 Violanthrene 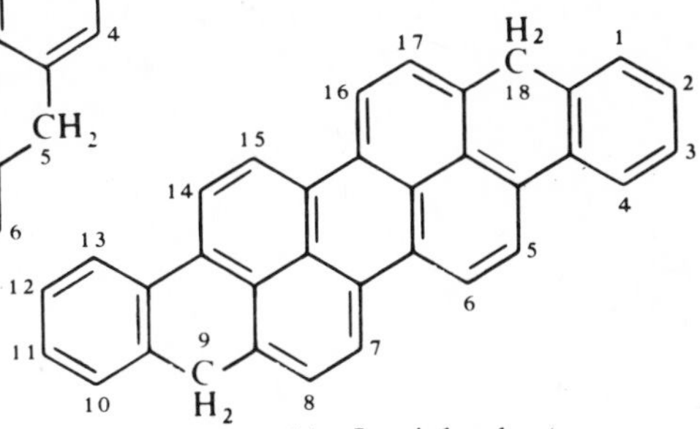   **14**   Isoviolanthrene

fused in a row). Pentacene, hexacene, and heptacene contain, respectively, five, six, and seven benzene rings fused in a row. Pentaphene, hexaphene, and heptaphene are dog-leg analogues.

THE FUSION (ANNELATION) METHOD FOR HYDROCARBONS

Aromatic polycyclic hydrocarbons not in the IUPAC list (*Table 5.2*) are named by fusion, for which also the IUPAC rules[1b] give the *CA* system. So many situations are to be covered that again the rules need consultation for complex cases, but the following is an outline of first principles.

(1)   The largest possible unit with a trivial name in the list is chosen as base component (in case of doubt, the component shown last in the list), and its sides are lettered *a* (italic) for side 1–2, and then *b, c*, etc., consecutively around the periphery, as in **15**. The complete name is then built up with an 'o' affix and square brackets as shown; the lowest available numbers (independently of substituents) are used for the fusion positions; and the numerals in the brackets must be in an order to correspond with 1,2 for side *a*, etc. The formula is then re-oriented, if necessary, to accord with rules (1)–(3) on p. 78 and *re-numbered* in accord with rule (4) as shown in formula **16**.
(2)   Case **17** is one of three-point fusion.
(3)   Designation within the square brackets is easier for benzo derivatives such as **18** and **19** (*see* p. 84).
(4)   Abbreviated names are used for the prefixes benzo-, naphtho-, anthra- (N.B. '-a'), phenanthro-, acenaphtho-, and perylo-, but not for others.
(5)   The final name applies to the compound with the maximum number of noncumulative double bonds; thus in indeno[1,2-*a*]indene (**20**) the characteristic CH_2 of indene has disappeared.

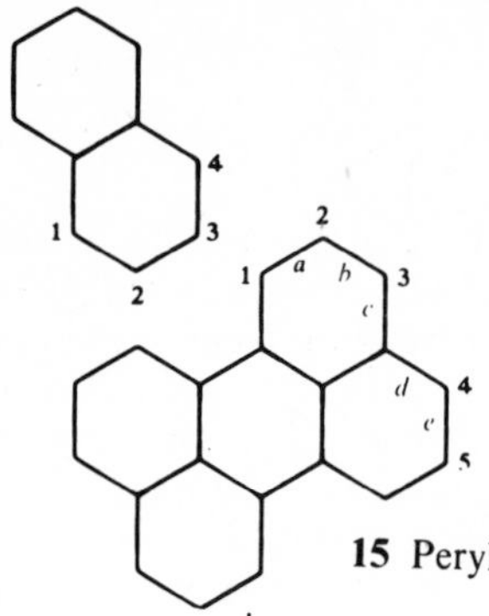

15 Perylene

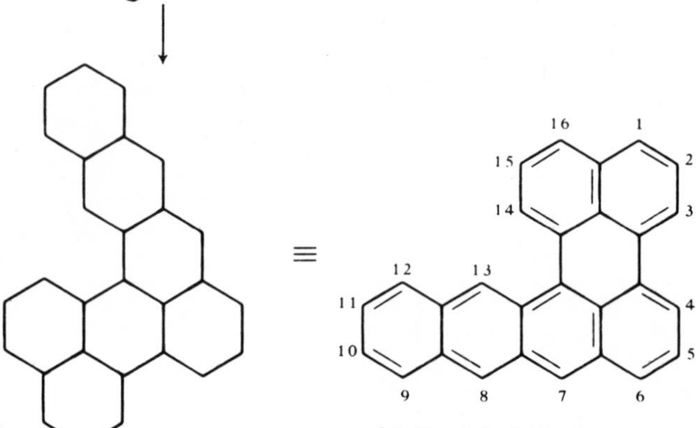

≡

16 Naphtho[2,3-*a*]perylene

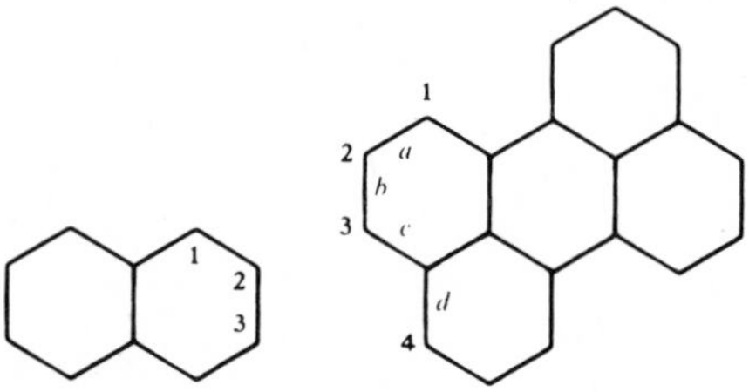

Perylene

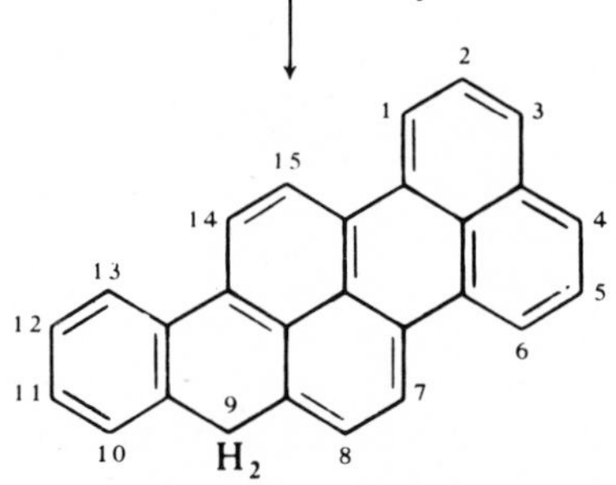

17 9*H*-Naphtho[1,2,3-*cd*]perylene

(6)   Reduction products are indicated by 'hydro-' prefixes (perhydro-for complete reduction).

(7)   Cycloalkene rings are denoted by prefixes 'cyclopenta-', 'cyclohepta-', 'cycloocta-', etc., which refer to the rings with the maximum number of noncumulative double bonds. Thus we arrive at the names shown for **21** and **22** [note in passing that the steroid nucleus **21** has the classical steroid (and not the general IUPAC) numbering; cf. p. 145]. The six-membered ring is named benzo also in cases such as **23**. When an unsaturated C_4 or C_3 ring is fused terminally to one of the systems in *Table 5.2*, an affix 'cyclobuta-' or 'cyclopropa-', respectively, may be used, as for

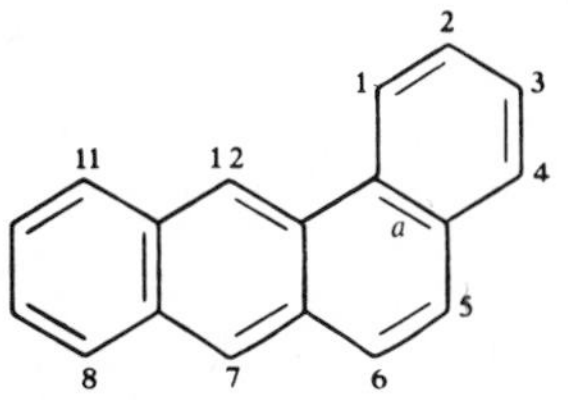

18 Benz[*a*]anthracene 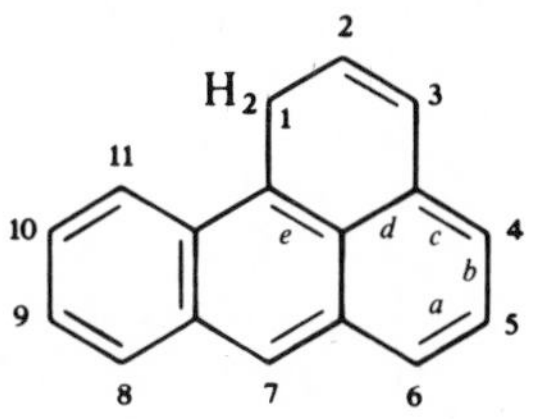   **19** 1*H*-Benz[*de*]anthracene

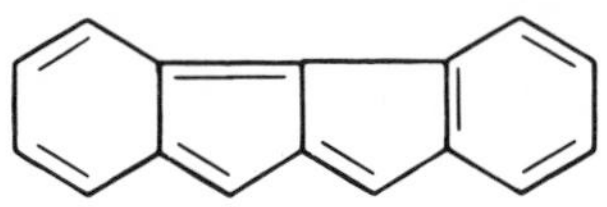

20 Indeno[1,2-*a*]indene

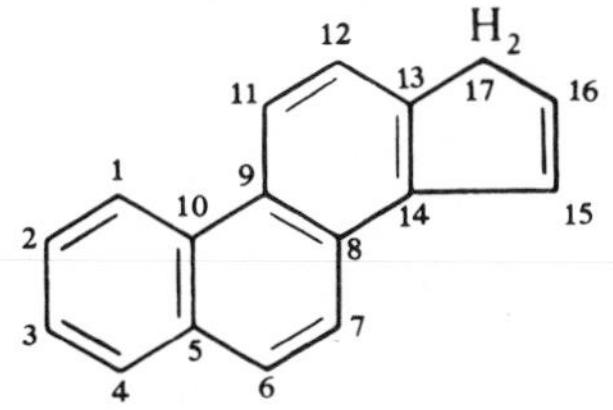

21 Cyclopenta[*a*]phenanthrene

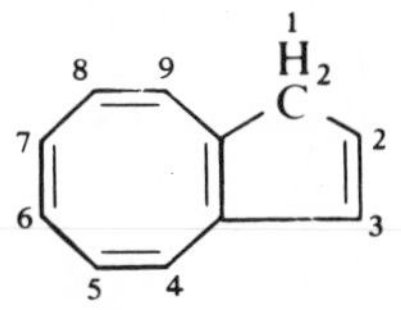

22 1*H*-Cyclopentacyclooctene

example in **24**. Actually such cases as **24** are rare; the IUPAC rule is so formulated that fusion principles are used only when two or more rings containing five or more ring members are fused together. All the compounds in *Table 5.2* are in accord with this, but in **25** there is only one ring with more than four ring members and such compounds must be named by the bicyclo method (see below). It may be emphasized that in names such as those for **21–24** the terminal '-ene' does not, as it usually does, mean one double bond, but maximum noncumulative double bonds throughout. Thus in spite of first appearance the 'cyclopenta-' portions of **21** and **22**, and the 'cyclobut-' portion in **24**

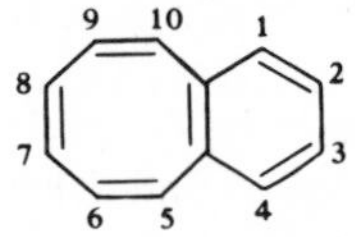

23 Benzocyclooctene

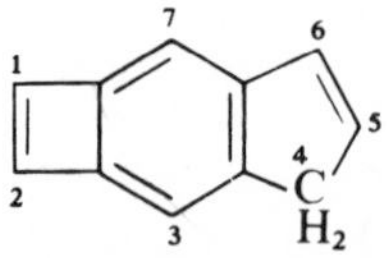

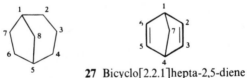

24 4*H*-Cyclobut[*f*]indene 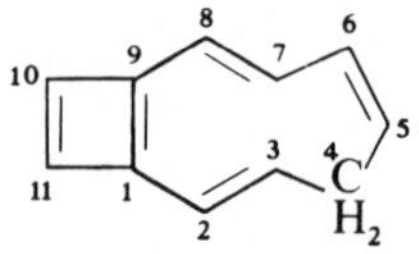   **25** 4*H*-Bicyclo[7.2.0]undecapentaene

contain double bonds, and '-cyclooctene' in **23** actually represents a
cyclooctatetraene derivative.

(8)   The 'o' indicating fusion is elided before a vowel, independently
of brackets, as in benz[*a*]anthracene **18**.

(9)   The reason why square brackets are used for fusion locants is
that the numerals do not correspond with those for the final numbering.

BICYCLO (ETC.) NOMENCLATURE

This system was devised primarily for compounds containing bridged
aliphatic rings, preferably saturated, where the resulting two or more
rings had two, or often more, atoms in common. Typical simple examples
are shown in formulas **26** and **27**. The method has, however, also to be

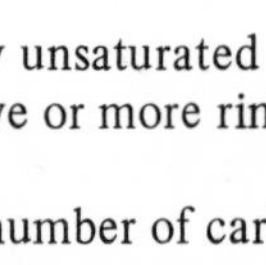

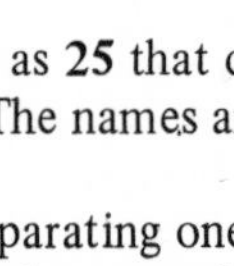

27 Bicyclo[2.2.1]hepta-2,5-diene

26 Bicyclo[3.2.1]octane

used for highly unsaturated systems such as **25** that do not contain two
rings having five or more ring members. The names are produced as
follows:

Count the number of carbon atoms separating one bridge end from
the other by the various routes and place these numbers, separated by
full stops, in descending order inside square brackets; in front of the
brackets place 'bicyclo' and after the brackets place the name of the
alkane containing the total number of carbon atoms (now including the
bridges), as shown under formulas **26** and **27**. One bridge may contain

zero carbon atoms, as in **25**. Numbering starts at one bridge end, goes to the other bridge end by the longest route, continues by the next longest route back to the first bridge end, and finally goes by the last (shortest) path to the second bridge end. Unsaturation is handled simply as in **27**. The method is often extended to tricyclic and more complicated bridge structures, and seems to be the only method of naming cage structures in general. Further rules necessary for such extensions are given by IUPAC[1c].

HYDROCARBON-BRIDGED (AROMATIC) SYSTEMS

Hydrocarbon bridges can be named by a quite different procedure that is most useful for bridges across aromatic systems. The bridge is denoted by means of the corresponding hydrocarbon molecule name (not the radical name), with a terminal '-o', e.g., methano $-CH_2-$, ethano $-CH_2 CH_2-$, or benzeno $-C_6 H_4-$, which is attached to the name of the ring system to be bridged. Compounds **28** and **29** provide two examples.

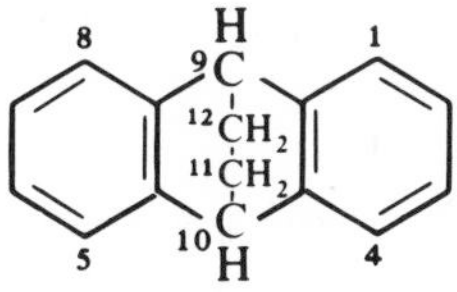

28 9,10-Dihydro-9,10-ethanoanthracene

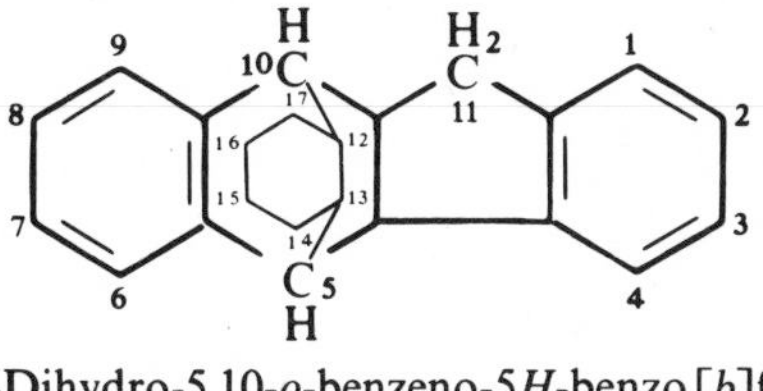

29 10,11-Dihydro-5,10-o-benzeno-5H-benzo[b]fluorene

Note two points even for such relatively simple systems: in both these examples, and in almost all other examples met in practice, the bridged ring is reduced — one hydrogen atom at each bridge end — so the names must then contain hydro prefixes. Secondly, the bridge is numbered starting from the end nearer the previous highest number. For further examples which can provide tricky problems of nomenclature the IUPAC rules can be consulted[1d].

SPIRANS

Spirans are structures in which two ring systems share one atom. As a result of IUPAC rules and general progress the various methods of naming them can now be simplified to the following two easy rules.

For two monocarbocycles in a spiran union the compound is named by placing 'spiro' before the name of the single large ring containing the same *total* number of carbon atoms; immediately after 'spiro' are placed numerals denoting the numbers of atoms linked to the spiro atom in each ring; these numbers start with the lower and are separated by a full stop (period) and placed in square brackets. Numbering of the complete compound starts from the carbon atom of the smaller terminal ring next to the spiro atom. Thus **30** is named spiro[4.5]deca-1,6-diene, and a simple extension of the method gives the name

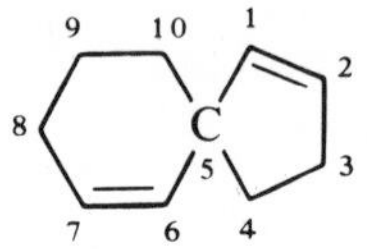

30 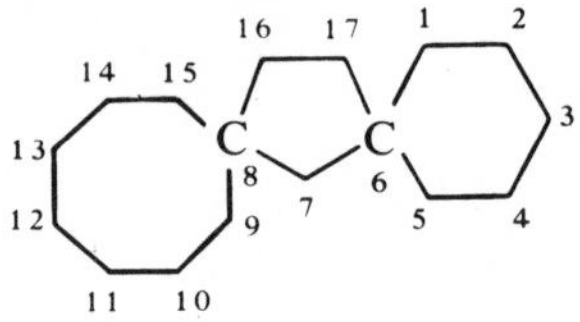   **31**

dispiro[5.1.7.2]heptadecane for **31**. When one (or more) component is a fused polycyclic system the different methods exemplified by 1,1'-spirobiindene for **32** and spiro[cyclopentane-1,1'-indene] for **33** are used.

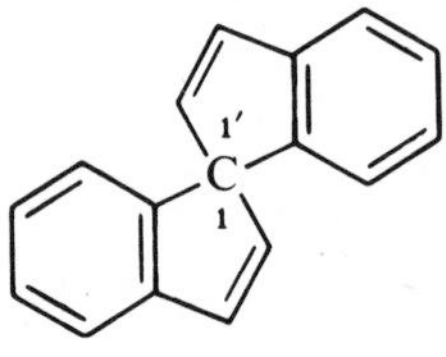

32 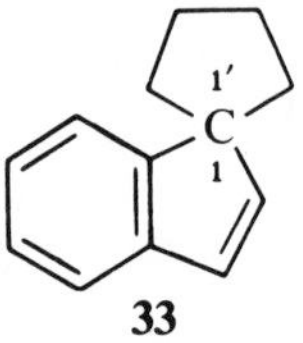   **33**

RINGS WITH SIDE CHAINS

A cyclic hydrocarbon with short side chains is normally treated as a substituted cyclic compound, e.g., hexamethylbenzene. A simple ring system with a long side chain is named by IUPAC rules as a substituted aliphatic hydrocarbon, e.g., 1-phenyldecane. Complex ring systems are normally treated as parents, even if they have long side chains, e.g., 1-dodecylpyrene. Under IUPAC rules choice is free for intermediate cases but *CA* always gives priority to the ring as parent. Unsaturation in the side chain causes IUPAC to tend to favour the side chain as parent, e.g., 1-(2-naphthyl)-1,3-hexadiene, which *CA* would call 2-(1,3-hexa-dienyl)naphthalene. Chains substituted by two or more cyclic radicals are most conveniently treated as parents, since, for example, 2-(2-naphthyl)-4-phenylhexane is 'simpler' than 2-(1-methyl-3-phenyl-pentyl)naphthalene. It must, however, always be remembered that the presence of a principal group may override any such considerations.

NEW CLASS NAMES

Newly invented names fall into two types: those that have been incor-porated into large groups such as carbohydrates and steroids whose systematization has been laboriously agreed among the relevant specialists (cf. Chapter 8); and those that do not clarify the structure at first glance, even to the specialist. There is, of course, also the lunatic fringe, which can be exemplified by the name barrelene for **34**; this name was given when interaction between the double bonds was postulated and illustrated as in **34**. Since it has been shown that this effect does not exist, it seems obvious that the not too difficult name bicyclo[2.2.2]octa-2,5,7-triene should be used, but such is the appeal of brevity that the misleading trivial name is still often employed.

34

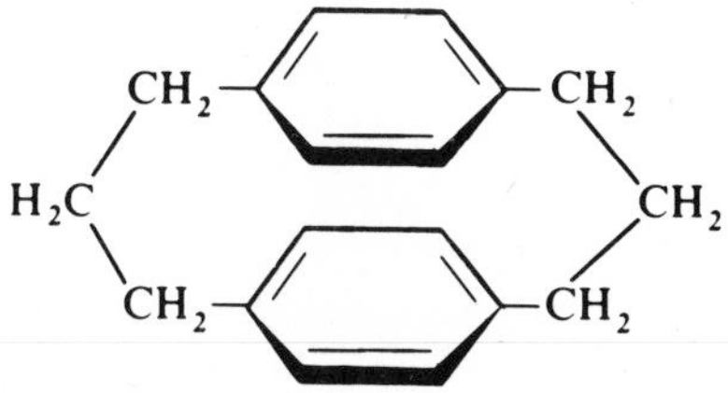

35

    An individual name sometimes becomes a class name as analogues are found to form an interesting new set of compounds. Such a case is the cyclophanes, now a generally accepted class name. A typical representative is **35**, for which a systematic name would be 1,1′:4,4′-bistrimethylenedibenzene but for which the convenient *p,p*′-[3.3]-cyclophane (*CA*: [3.3]paracyclophane) was invented. Obvious small changes allow for other linking alkylene groups and for *ortho* and *meta* linkage, but differing types of variant can be devised for cases where the benzene rings are replaced by other nuclei, when the alkylene chains contain heteroatoms, where more than two rings are linked in the way shown, or for more than two linking chains.

There is need for systematization of this growing field, and it is currently being considered by an IUPAC working group.

Heterocyclic Compounds

Of all classes of compound the heterocyclic series presents the greatest variety of structure and thus the most complicated nomenclature. The following is a general outline.

TRIVIAL NAMES

There is a profusion of trivial or semitrivial names, even when the alkaloids are excluded. The IUPAC rules list names of 47 heterocyclic skeletons that may be used in fusion operations (*see Table 5.3*) and of 14 hydrogenated systems that should not be used in fusion operations (*see Table 5.4*). Throughout the heterocyclic series a heteroatom at a ring junction is numbered sequentially as in quinolizine (no. 25 in *Table 5.3*). The alternative names labelled (*CA*) are those currently used in *CA* indexes, which explicitly designate all indicated hydrogen instead of leaving it implied in the commonest isomer as some IUPAC names do.

EXTENDED HANTZSCH–WIDMAN SYSTEM

Except for trivial names, this nomenclature has the widest use. Syllables ending in '-a' (elided before a vowel) denote the heteroatoms (*see Table 5.5*) and are followed by other syllables to denote ring size (*see Table 5.6*, p. 95); the latter are in some cases modified to denote the state of hydrogenation of the ring, but in others this always requires hydro prefixes to be used. Most of the suffixes are derived by selection of letters from the appropriate numeral, 'ir' from *tri* (for three-membered rings), 'et' from te*t*ra, 'ep' from he*p*ta, 'oc' from *oc*ta, 'on' from n*on*a, and 'ec' from d*ec*a; but 'ole' for a five-membered and 'in' for a six-membered ring are from the original Hantzsch–Widman endings[2]. For six- or higher-membered rings a terminal 'ine' distinguishes unsaturated nitrogenous rings; '-idine' is used as ending for fully reduced three- to five-membered rings, and '-ane' for all fully reduced non-nitrogenous rings; partly reduced four- and five-membered rings have the endings shown in *Table 5.7* (the '-oline' ending is from Hantzsch–Widman).

When more than one heteroatom is present in a ring they are cited in descending Group number of the Periodic Table and in increasing

[*continued on p. 95*

Table 5.3 TRIVIAL AND SEMITRIVIAL NAMES OF HETEROCYCLES, IN ORDER OF 'SENIORITY' WHOSE NAMES ARE APPROVED BY IUPAC AND/OR *CA* FOR USE IN FUSION NOMENCLATURE*†

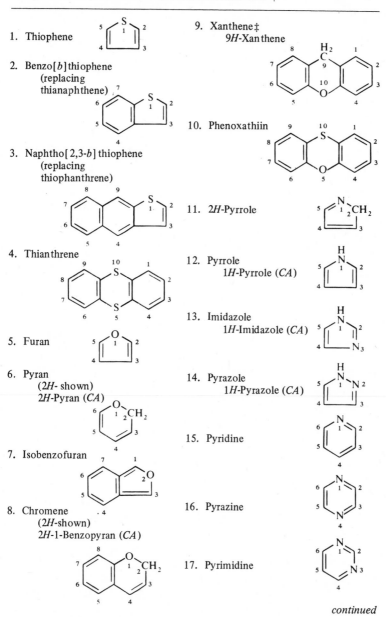

1. Thiophene

2. Benzo[*b*]thiophene (replacing thianaphthene)

3. Naphtho[2,3-*b*]thiophene (replacing thiophanthrene)

4. Thianthrene

5. Furan

6. Pyran (2*H*- shown) 2*H*-Pyran (*CA*)

7. Isobenzofuran

8. Chromene (2*H*-shown) 2*H*-1-Benzopyran (*CA*)

9. Xanthene‡ 9*H*-Xanthene

10. Phenoxathiin

11. 2*H*-Pyrrole

12. Pyrrole 1*H*-Pyrrole (*CA*)

13. Imidazole 1*H*-Imidazole (*CA*)

14. Pyrazole 1*H*-Pyrazole (*CA*)

15. Pyridine

16. Pyrazine

17. Pyrimidine

continued

Table 5.3 (*continued*)

18. Pyridazine

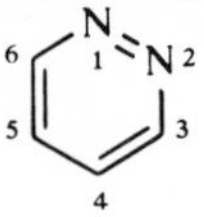

19. Indolizine

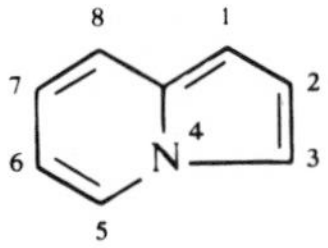

20. Isoindole
    2*H*-Isoindole (*CA*)

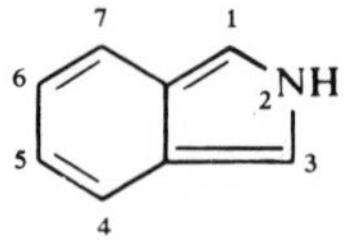

21. 3*H*-Indole

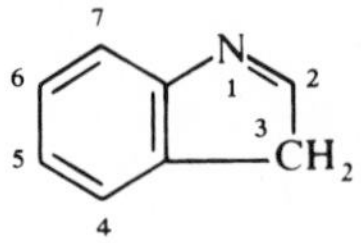

22. Indole
    1*H*-Indole (*CA*)

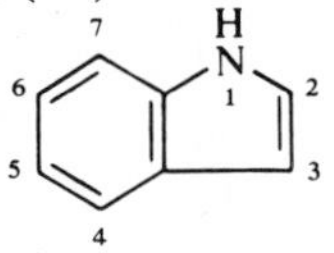

23. 1*H*-Indazole

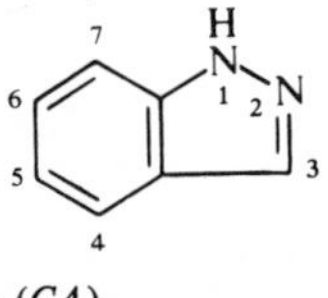

24. Purine‡
    7*H*-Purine (*CA*)

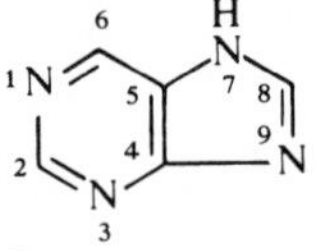

25. 4*H*-Quinolizine

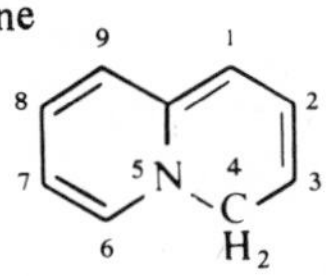

26. Isoquinoline

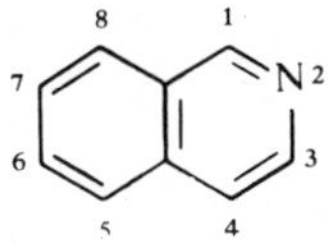

27. Quinoline

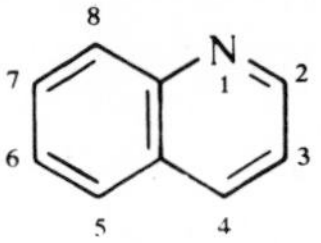

28. Phthalazine

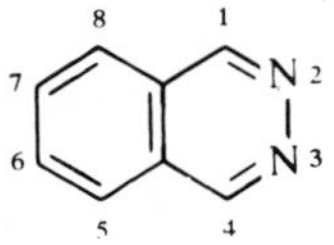

29. Naphthyridine
    (1,8-shown)

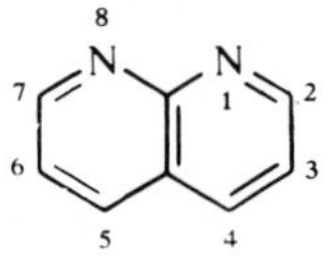

30. Quinoxaline

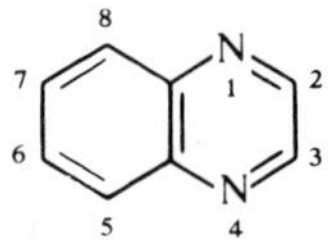

31. Quinazoline

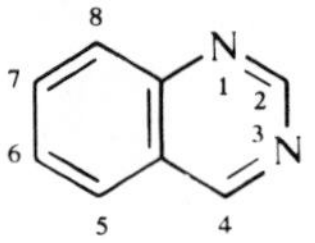

32. Cinnoline

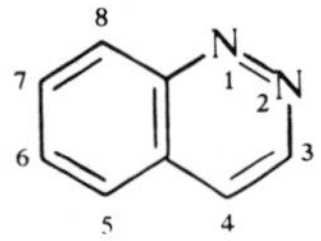

33. Pteridine

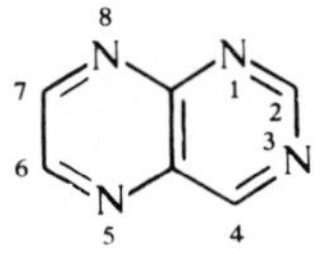

34. 4a*H*-Carbazole‡

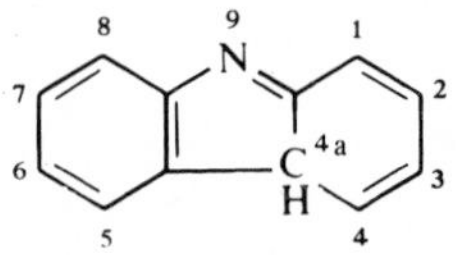

35. Carbazole‡
    9*H*-Carbazole (*CA*)

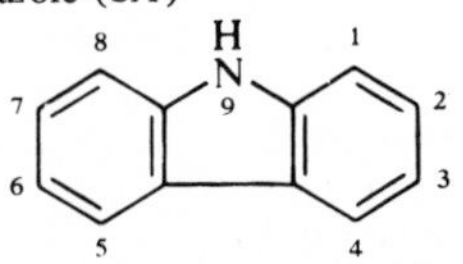

continued

Table 5.3 (*continued*)

36. β-Carboline
    9*H*-Pyrido[3,4-*b*]-
    indole (*CA*)

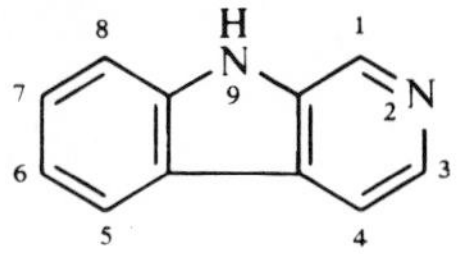

37. Phenanthridine

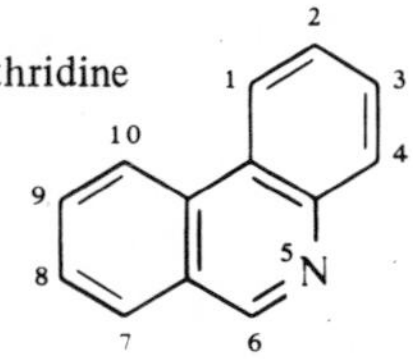

38. Acridine‡

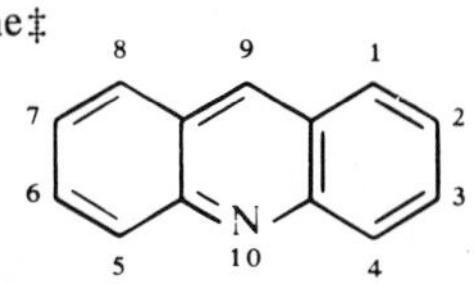

39. Perimidine
    1*H*-Perimidine (*CA*)

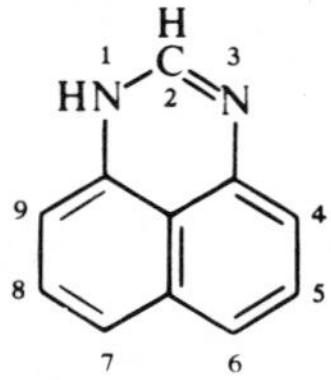

40. Phenanthroline
    (1,7-shown)

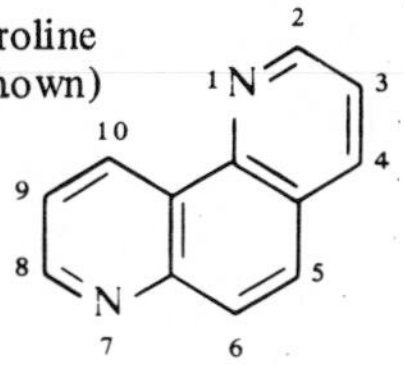

41. Phenazine

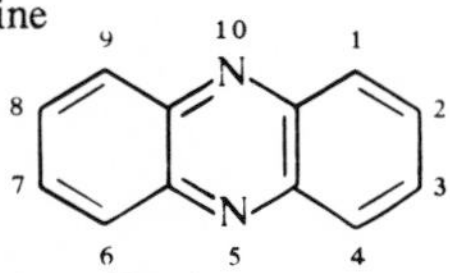

42. Phenarsazine

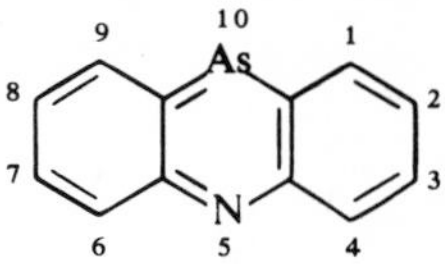

43. Isothiazole

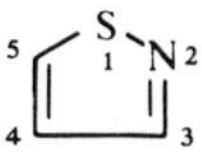

44. Phenothiazine
    10*H*-Phenothiazine (*CA*)

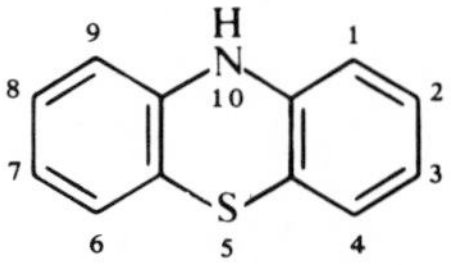

45. Isoxazole

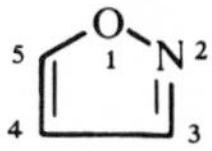

46. Furazan

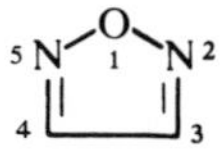

47. Phenoxazine
    10*H*-Phenoxazine (*CA*)

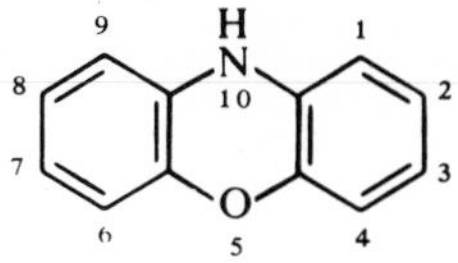

---

*For meaning of seniority, *see* p. 79.
†In tentative rules (as Appendix 4 to IUPAC section D; see page 179) items 2
  and 3 are deleted and 28 further compounds are added to this list.
‡Denotes exception to systematic numbering.

Table 5.4 TRIVIAL AND SEMITRIVIAL NAMES* OF REDUCED HETEROCYCLIC SYSTEMS, NOT TO BE USED IN FUSION NOMENCLATURE

1. Isochroman

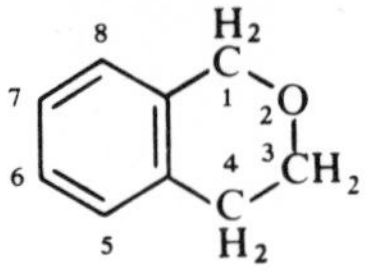

2. Chroman

3. Pyrrolidine

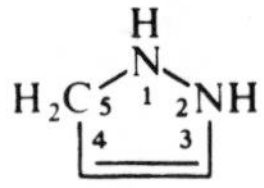

4. Pyrroline (2- shown†)

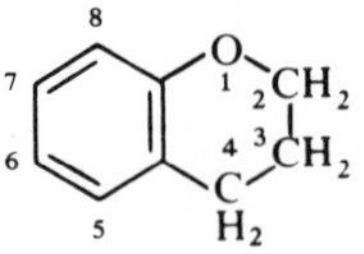

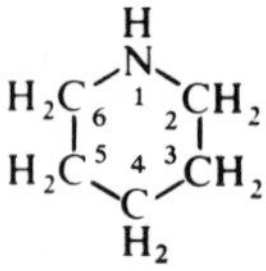

5. Imidazolidine

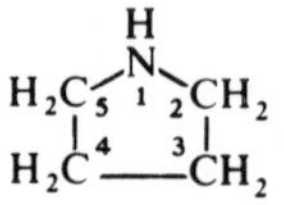

6. Imidazoline (2-shown†)

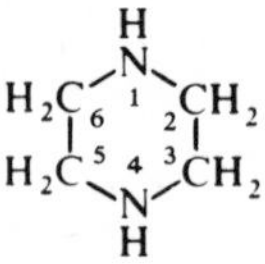

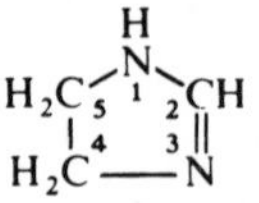

7. Pyrazolidine

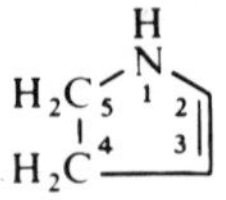

8. Pyrazoline (3-shown†)

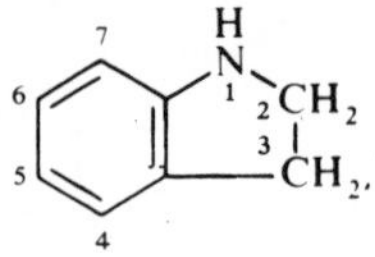

9. Piperidine

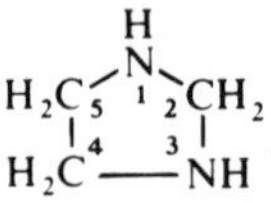

10. Piperazine

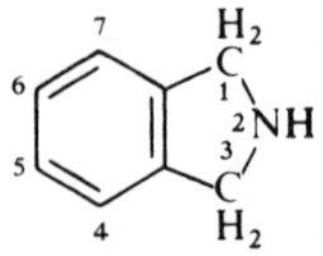

11. Indoline

12. Isoindoline

13. Quinuclidine

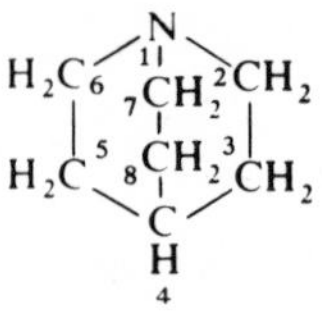

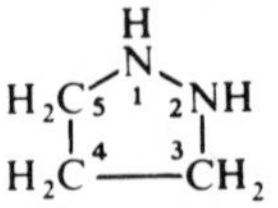

14. Morpholine

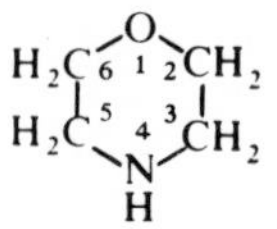

*These are IUPAC names; in some cases *CA* uses a more systematic name.
†The numeral denotes the position of the double bond.

atomic number, e.g., 'oxathia', 'thiaza', 'oxaza' (note the elision of 'a' from 'thiaaza' and 'oxaaza').

Locants precede the names thus produced. Subject to being as low as possible, they must give the lowest permissible number to the hetero-atom given first in *Table 5.5* and then to other heteroatoms; they are arranged in the same order as the heteroatoms. As examples, one

Table 5.5 PREFIXES TO DENOTE HETEROATOMS IN RINGS

Element	Valence	Prefix	Element	Valence	Prefix
Oxygen	II	oxa	Antimony	III	stiba*
Sulfur	II	thia	Bismuth	III	bismutha
Selenium	II	selena	Silicon	IV	sila
Tellurium	II	tellura	Germanium	IV	germa
Nitrogen	III	aza	Tin	IV	stanna
Phosphorus	III	phospha*	Lead	IV	plumba
Arsenic	III	arsa*	Boron	III	bora
			Mercury	II	mercura

*When immediately followed by '-in' or '-ine', 'phospha-' should be replaced by 'phosphor-', 'arsa-' should be replaced by 'arsen-', and 'stiba-' should be replaced by 'antimon-'. In addition, the saturated six-member rings corresponding to phosphorin and arsenin are named phosphorinane and arsenane.

Table 5.6 SUFFIXES USED TO DENOTE RING SIZE AND STATE OF REDUCTION OF THE RING IN NOMENCLATURE OF HETEROCYCLES BY THE EXTENDED HANTZSCH–WIDMAN SYSTEM

Number of members in the ring	Rings containing nitrogen		Rings containing no nitrogen	
	Unsaturation*	Saturation	Unsaturation*	Saturation
3	-irine	-iridine	-irene	-irane
4	-ete	-etidine	-ete	-etane
5	-ole	-olidine	-ole	-olane
6	-ine†	‡	-in†	-ane§
7	-epine	‡	-epin	-epane
8	-ocine	‡	-ocin	-ocane
9	-onine	‡	-onin	-onane
10	-ecine	‡	-ecin	-ecane

*Corresponding to the maximum number of noncumulative double bonds, the hetero elements having the normal valences shown in *Table 5.5*.
†For phosphorus, arsenic, and antimony, see the special provisions of *Table 5.5*.
‡Expressed by using 'hexahydro-' or 'octahydro-' prefixes, as appropriate, or else 'perhydro-'.
§Not applicable to silicon, germanium, tin, and lead. In this case, 'perhydro-' is prefixed to the name of the corresponding unsaturated compound.

Table 5.7 SUFFIXES DENOTING RING SIZE FOR SOME PARTLY HYDROGENATED RINGS

Number of members of the partly saturated rings	Rings containing nitrogen	Rings containing no nitrogen
4	-etine	-etine
5	-oline	-olene

Examples:

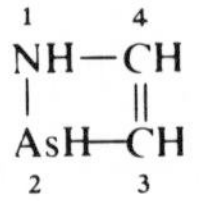

NH—CH¹ ⁴
| ‖
AsH—CH
²   ³

Δ^3-1,2-Azarsetine*

HC==CH
H₂C⌐SiH₂⌐CH₂

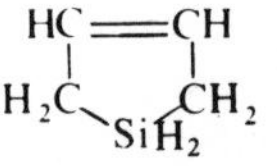

Δ^3-Silolene

*As exceptions, Greek capital delta (Δ), followed by superscript locant(s), is used to denote a double bond in a compound named according to this system if its name is preceded by locants for heteroatoms.

obtains 1,3-thiazole for **36**, 1,2,4-triazine for **37**, and 1,2,4-thiazaphosphole for **38**. (1,3-Thiazole is usually shortened to thiazole.)

Indicated hydrogen often must be cited as for carbocycles, e.g., 6H-1,2,5-thiadiazine (**39**).

36 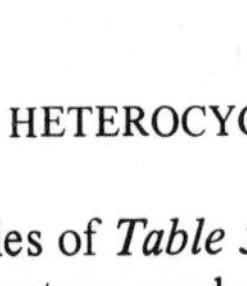   **37**   **38**   **39**

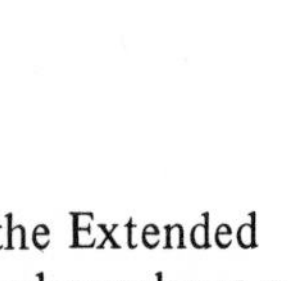

FUSION NAMES FOR HETEROCYCLES

Names of heterocycles of *Table 5.3* or those formed by the Extended Hantzsch–Widman system can be 'fused' with those of hydrocarbons or other heterocyclic systems by the methods outlined on pp. 83–86 for hydrocarbons. Naturally some special points arise.

The following abbreviations are used in fusion names of heterocycles: furo, imidazo, pyrido, quino, and thieno, but not others.

The heteroatoms in the complete structure receive the lowest numbers permissible when the structure is oriented as for carbocycles; *but* heteroatoms at ring junction (valley) positions are numbered serially without the use of a, b, etc. Formula **40** for pyrano[3,4-b]quinolizine illustrates both points.

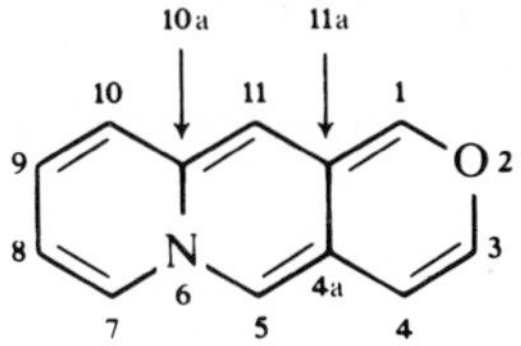

40

Square brackets are used as for carbocycles to distinguish locants that refer to the unfused component and not to the final compound. Examples are anthra[2,1-*d*]thiazole **41** and pyrrolo[2,3-*c*]carbazole **42**.

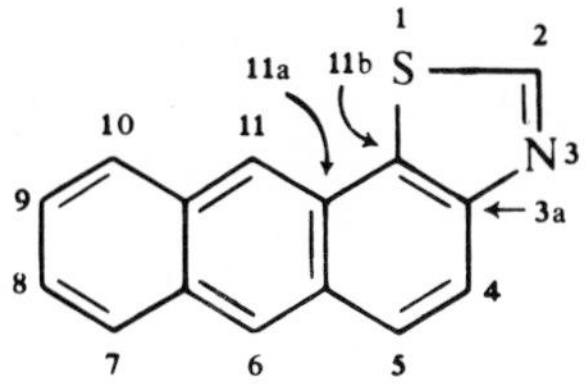

41

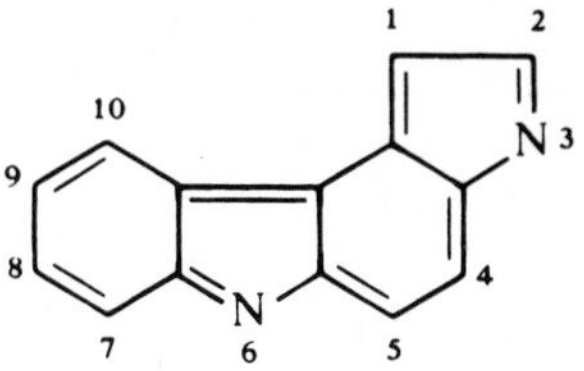

42

Compound **42** serves also to illustrate the loss of two hydrogen atoms (one from the pyrrole and one from the carbazole component) to obtain maximum noncumulative unsaturation.

Many intricate problems arise in this field, mainly in connection with choosing the components, and with numbering, and workers in this field should familiarize themselves with the details of Section B of the IUPAC rules[1] and with the *Ring Index*[3] or its successor, the *Parent Compound Handbook*[4]. A comprehensive discussion of the nomenclature of heterocyclic compounds, reprinting the IUPAC rules, is also available[5].

However, there is one welcome simplification, when a benzene ring is one fusion component to the Hantzsch–Widman type of name (incidentally a common type of compound): here names such as 1,2-benzoxathiin and 2,1-benzoxathiin are adequate to distinguish **43** from **44**, respectively, and there are many similar cases.

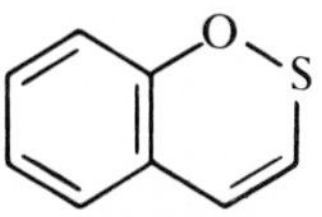

43

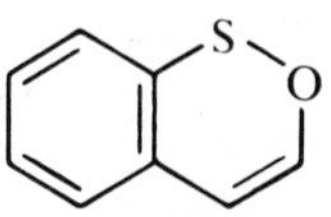

44

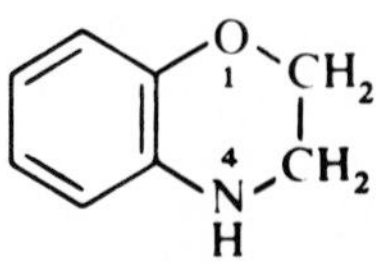

45

All complete fused structures containing less than the maximum number of double bonds must have 'hydro-' prefixes in the name, and these prefixes receive the lowest possible numbers *after* allowance for indicated hydrogen: the name for **45** is 3,4-dihydro-2*H*-1,4-benzoxazine.

REPLACEMENT NOMENCLATURE

The terms denoting heteroatoms listed in *Table 5.5* for the Extended Hantzsch–Widman names can also be used for general replacement nomenclature. When prefixed to the name of a hydrocarbon they denote replacement of a carbon atom of that compound by the stated heteroatom. This gives names that often prove most useful, particularly for compounds containing several types of heteroatomic ring members or complex ring systems. An example is 8*H*-7-thia-1,9-diazaphenanthrene **46**. It will be seen that the numbering of the hydrocarbon is retained but, so far as that retention permits, the heteroatoms have lowest numbers.

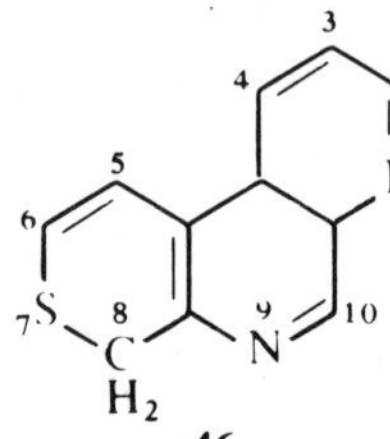

46

This nomenclature must be applied to the hydrocarbon parent, not to a less hetero-substituted system — to avoid plurality of names such as, for instance, 4*H*-3-thia-7-azaphenanthridine for **46**.

Bicyclo, tricyclo, etc., names are also a very fruitful field for replacement nomenclature.

HETERO BRIDGES

Hetero bridges, particularly those across aromatic or complex systems, can be usefully named by the 'epi' system. The prefix 'epi-' ('ep-' before a vowel) is followed by the name of the bridging *radical* (contrast the method for bridging hydrocarbons, p. 87). One thus obtains epoxy- for a bridging –O–, 'epidioxy-' for –O–O–, 'epithio-' for –S–, 'epimino-' for –NH–, and so on. Thus compound **47** can be called 9,10-dihydro-9,10-epoxyanthracene and **48** 1,9-dihydro-1,9-epidioxyphenanthrene. Note the 'dihydro' prefixes, as for bridging hydrocarbons, also that 'epi' terms

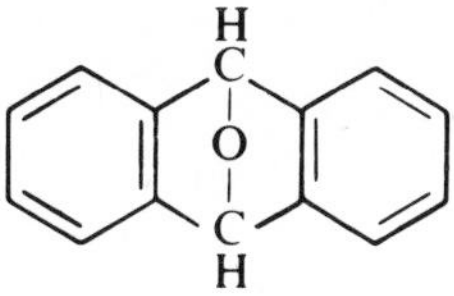

47 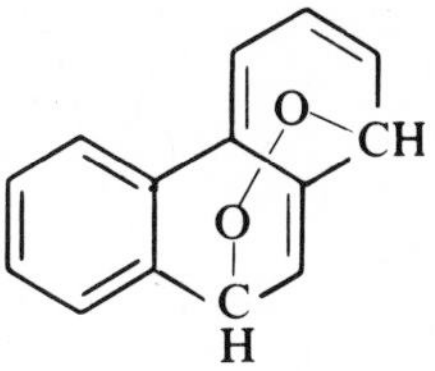 48

are considered to be part of the main structure — i.e., nondetachable —
and are not alphabetized among the normal prefixes.

REFERENCES

1. *IUPAC Nomenclature of Organic Chemistry, Definitive Rules for: Section A.
   Hydrocarbons; Section B. Fundamental Heterocyclic Systems; Section C.
   Characteristic Groups Containing Carbon, Hydrogen, Oxygen, Nitrogen,
   Halogen, Sulfur, Selenium and/or Tellurium*, 1969. A, B 3rd Ed.; C 2nd Ed.,
   Butterworths, London (1971); now available from Pergamon Press, Oxford.
   (a) pp. 8–11, 13–15; (b) pp. 22–29; (c) pp. 32–34; (d) pp. 35–37
2. HANTZSCH, A. and WEBER, J.H., *Ber. Deut. Chem. Ges.*, 20, 3119 (1887);
   WIDMAN, O., *J. Prakt. Chem.* (2), 38, 185 (1888)
3. PATTERSON, A.M., CAPELL, L.T., and WALKER, D.F., *The Ring Index*,
   American Chemical Society, Washington, D.C., 1960; *Supplement 1*, 1963;
   Supplement 2, 1964; *Supplement 3*, 1965
4. *Parent Compound Handbook*, Chemical Abstracts Service (1977)
5. MCNAUGHT, A.D., *Adv. Heterocycl. Chem.*, 20, 175 (1976)

6

Organic: Some Special Features and Functional Groups

Radicals (As Substituents) From Hydrocarbon and Heterocyclic Molecules

INTRODUCTION

The naming of radicals is an area in which many older names approved by IUPAC and still used by most chemists have been replaced in *CA* indexes by others considered more systematic. Thus both kinds have to be discussed here. It appears likely that the logically simpler (but not shorter) *CA* names will have increasing appeal.

ACYCLIC

As is well known, univalent radical names are derived from those of alkanes, alkenes, or alkynes by changing the -ane, -ene, or -yne ending to -yl, -enyl, or -ynyl respectively, e.g., methyl, 2-butenyl. The following nonsystematic names are authorized by IUPAC but no longer used by *CA*: $CH_2=CH-$ vinyl, $CH_2=CHCH_2-$ allyl.

Similarly IUPAC but not *CA* permits the naming of a few unsubstituted radicals with the iso prefix [for C_3-C_6, e.g. $(CH_3)_2CH-$ isopropyl)], *sec-* [in $CH_3CH_2CH(CH_3)-$ *sec*-butyl only], neo [for $(CH_3)_3CCH_2-$ neopentyl, only], and *tert-* [$(CH_3)_3C-$ *tert*-butyl; $CH_3CH_2C(CH_3)_2-$ *tert*-pentyl].

To denote the position of substituents within radicals, the point of attachment receives number 1, and the most unsaturated, or most substituted, or longest (in that order) chain starting from that atom numbered 1 is regarded by IUPAC as the main (parent) chain*. Thus for instance heptane C_7H_{16} gives radicals $CH_3(CH_2)_5CH_2 -$ heptyl, $CH_3(CH_2)_4CH(CH_3)-$ 1-methylhexyl, $CH_3(CH_2)_3CH(CH_3)CH_2 -$ 2-methylhexyl, etc. Similarly $CH_3CH_2CH_2CH(C_2H_5)CH_2 -$ is 2-ethylpentyl, but $CH_3CH_2CH_2CH(CH=CH_2)CH_2 -$ is 2-propyl-3-butenyl (or 2-ethenylpentyl in *CA*). The IUPAC trivial name isopropenyl for $CH_2=C(CH_3)-$ is a convenient exception. Note that points of attachment numbered other than 1 are *not* permitted; thus 2-propyl should be called either isopropyl or 1-methylethyl.

Names of radicals $CH_3(CH_2)_xCH=$ end in -ylidene (in place of -ane) and the rare $CH_3(CH_2)_xC\equiv$ are analogously called alkylidyne radicals.

According to IUPAC, $-CH_2 -$ is methylene and $CH_2=$ has the same name; $-CH_2CH_2 -$ is called ethylene, and $-CH(CH_3)CH_2 -$ propylene, while the higher unbranched diradicals $-CH_2(CH_2)_x -$ are named as polymethylenes, e.g., $-CH_2(CH_2)_3CH_2 -$ pentamethylene. (N.B. The hydrocarbon $CH_3CH=CH_2$ is systematically propene, not propylene.) Of these names *CA* uses only methylene, and applies the ending -diyl to designate the rest, e.g., $-CH_2CH_2CH_2 -$ 1,3-propanediyl; $-CH(CH_3)CH_2 -$ 1-methyl-1,2-ethanediyl (not 1,2-propanediyl, which would not give priority to the points of attachment for defining the parent alkane).

Unsaturated radicals are named analogously, e.g., $-CH_2CH=CHCH_2 -$ 2-butenylene (IUPAC) or 2-butene-1,4-diyl (*CA*).

RADICALS FROM CYCLIC SYSTEMS

Radicals formed from carbomonocycles are named analogously to those from alkanes, alkenes, etc. The generic name aryl is similarly derived from arene. For all other radicals formed from cyclic molecules, there is a general rule that they are named by *adding* '-yl', '-ylene' (or '-diyl'), '-ylidene', '-triyl', etc., as appropriate, to the name of the compound (with elision of any terminal '-e' before '-y'). Examples are indenyl, indenediyl, azulenyl, carbazolyl, pteridinyl, and isoxazolyl. *Table 6.1* is a list (page 102) showing some exceptions, made mostly by IUPAC but no longer by *CA*.

NUMBERING

In all cases, locants for radical endings are as low as permissible by any fixed numbering of the system; they have priority over locants for

*In *CA*, however, the longest chain has priority.

Table 6.1   RADICALS FORMED FROM SOME SIMPLE CYCLIC SYSTEMS

IUPAC name	CA name	IUPAC name	CA name
Anthryl	Anthracenyl	Phenethyl	2-Phenylethyl
Benzhydryl *or*	Diphenylmethyl	Phenyl	Phenyl
Diphenylmethyl		Phenylene	Phenylene
Benzyl	Phenylmethyl	*x*-Piperidyl	*x*-Piperidinyl
Benzylidene	Phenylmethylene	Piperidino	1-Piperidinyl
Cinnamyl	3-Phenyl-2-	Pyridyl	Pyridinyl
	propenyl	Quinolyl	Quinolinyl
Furyl	Furanyl	Styryl	2-Phenylethenyl
Isoquinolyl	Isoquinolinyl	Thenyl	2-Thienylmethyl
Mesityl	2,4,6-Trimethyl-	Thienyl	Thienyl
	phenyl	Tolyl	Methylphenyl
Naphthyl	Naphthalenyl	Trityl	Triphenylmethyl
Phenanthryl	Phenanthrenyl	Xylyl	Dimethylphenyl

unsaturation and all substituents. For example, in compound **1**, the naphthalene ring, being 'senior' to cyclohexene, is named as parent, so that the latter must be cited as a substituent. This gives the numbering shown and the name as 7-(6-carboxy-2-methyl-2-cyclohexen-1-yl)-2-naphthoic acid. (Note that the direction of numbering of the cyclohexene ring gives preference to the double bond over the carboxyl group since the latter is not a principal group here.)

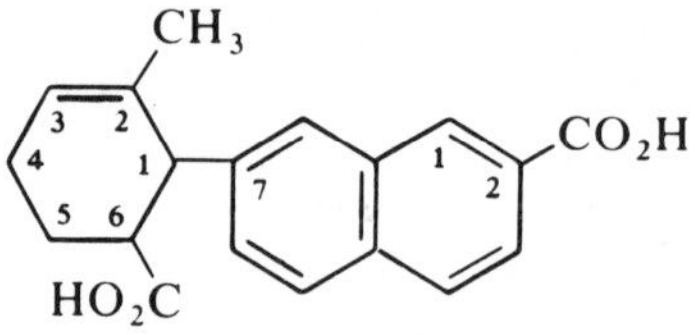

1

Free Radicals

Free radicals from alkanes or cyclic structures have the same names as for substituent groups, e.g., $CH_3\cdot$ methyl, $C_6H_5\cdot$ phenyl. The electron-deficient species $:CH_2$ and its substitution products may be named either as methylenes or as carbenes, e.g., $Cl_2C:$ dichloromethylene or dichlorocarbene. Collectively they are called carbenes.

The -yl ending is used for free radicals also when the group name may end in -y, e.g., $CH_3O\cdot$ methoxyl (not methoxy); and amine free radical names also end in '-yl', e.g., $CH_3NH\cdot$ methylaminyl, and 1-piperidinyl (not piperidino). Specialists may note IUPAC names $CH_3N\cdot$

methylaminylene and its analogues, also hydrazyl for $NH_2NH\cdot$, but methylnitrene (by analogy with carbene) is a commoner name for CH_3N, whether the triplet or the singlet state is involved.

Ions

Anions of acids have names formed by replacing '-ic acid' by '-ate', or '-ous acid' by '-ite' (*see* pp. 18, 19, 111).

Anions formed by removing a proton (or protons) from carbon are often called carbanions. They are named by adding '-ide' (or '-diide', etc.) to the name of the parent compound (with elision of 'e' before a vowel). For example, $CH_3CH_2CH_2CH_2^-$ 1-butanide, $C_6H_5^-$ benzenide, and 1,4-dihydro-1,4-naphthalenediide for **2**.

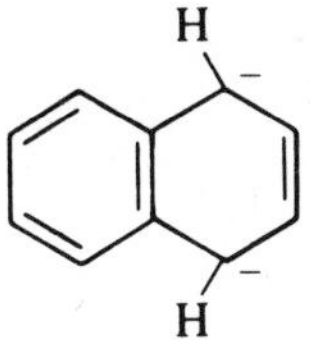

2

Cation nomenclature is not simple, except that the ending -ium (with elision of a preceding vowel) is used in all normal cases. We may first note ammonium H_4N^+, sulfonium H_3S^+ and similar heteroatomic cations and their derivatives (cf. p. 23). Next, for cations derived from compounds by addition of a proton the '-ium' is added to the systematic or trivial name, with addition of any necessary locants. Typical examples are anilinium $C_6H_5NH_3^+$, 1-methylhydrazinium **3** ($N^+ = 1$); and 9aH-quinolizinium **4**.

$$\overset{+}{NH_2}NH_2CH_3$$

3

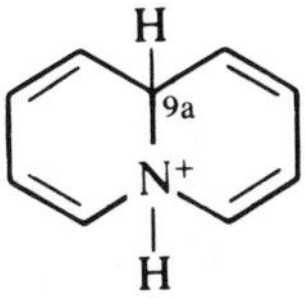

4

Uronium is an exception.

$$OC(NH_2)(\overset{+}{NH_3}) \rightleftharpoons HO-C(NH_2)=\overset{+}{NH_2}$$

Uronium

Cations formally derivable by addition of a proton to an unsaturated compound are named from the latter, e.g., ethenium $CH_3CH_2^+$, benzenium $C_6H_7^+$, and allenium as shown below.

$$H^+$$

$$\overline{}$$

$$CH_2{=}C{=}CH_2$$

Allenium

Cations formally derivable by loss of an electron from a free radical may be named from the radical, e.g., acetylium CH_3CO^+, phenylium $C_6H_5^+$; this method is currently preferred by *CA*. Cations formally derivable by addition of a proton to an unsaturated compound may be named from the latter, e.g., ethenium $CH_3CH_2^+$, which by the radical method is called ethylium. Such cationic structures may involve bond delocalization, as in benzenium $C_6H_7^+$ (dihydrophenylium) and the most probable protonated form of allene, 2-allenium $[CH_2{\cdots}CH{\cdots}CH_2]^+$ (propenylium). Another simple method of naming such cations is merely to name the radical and add the separate word 'cation', e.g., ethyl cation, allyl cation (the latter for the propenylium·ion shown above). Although sometimes used, the term 'carbonium' has no IUPAC sanction and is logically objectionable. It is even possible to add a proton to an alkane, to produce strange ions such as CH_5^+; pending the issuance of applicable official rules, they are best named merely by description, such as 'protonated methane'. The class name carbocation is also unofficial but useful for referring to compounds containing positively charged carbon atoms.

Radical cations have names ending in '-iumyl', e.g., $C_6H_6^{+\cdot}$ benzeniumyl, quinoliniumyl $C_9H_7N^{+\cdot}$, dimethylsulfoniumyl $(CH_3)_2S^{+\cdot}$.

For prefixes, '-ium' is changed to '-io-', as in 5, 1-methyl-4-(trimethyl-ammonio)quinolinium dichloride. In replacement names, a charged

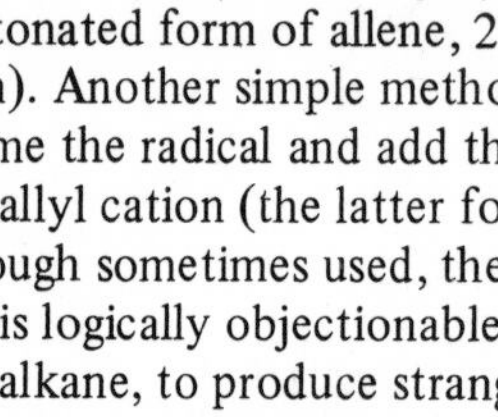

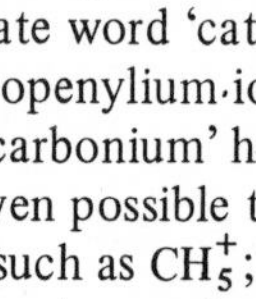

heteroatom is cited by using, e.g., '-oxonia-', '-azonia-', '-thionia-' instead of 'oxa-', '-aza-', '-thia-'. This proves very useful in bicyclo nomenclature, as in 6, the 1-ethyl-1-azoniabicyclo[3.3.0]octane ion.

For further examples, the IUPAC rules[1a] should be consulted.

Halo (Halogeno) Compounds

Substitutive nomenclature, with fluoro, chloro, bromo, and iodo prefixes for halogen substituents, as in 2-chloroquinoline, 2,3-dichloropropionic acid, or 1-bromo-6-iodohexane, is widely used and convenient, but for very simple compounds radicofunctional nomenclature, such as in methyl iodide CH_3I, benzylidene dichloride (not benzal dichloride) $C_6H_5CHCl_2$, and ethylene dichloride $ClCH_2CH_2Cl$, still has its attractions as emphasizing the main or sole site of reactivity. Also additive nomenclature is sometimes informative when two halogen atoms have been or can be considered to have been added to a double bond, as in styrene dibromide $C_6H_5CHBrCH_2Br$.

When every hydrogen atom of a parent compound is replaced by the *same kind* of halogen atom a prefix 'per-' may be used to denote complete replacement. Note, however, the difference between perfluoro(decahydro-2-methyl-1-naphthoic acid) $CF_3C_{10}F_{16}CO_2H$ and hexadecafluorodecahydro-2-methyl-1-naphthoic acid $CH_3C_{10}F_{16}CO_2H$ as well as decahydro-2-(trifluoromethyl)-1-naphthoic acid $CF_3C_{10}H_{16}CO_2H$.

Probably because of their brevity and long use, haloform names, e.g., fluoroform, chloroform, for $CH(Hal)_3$ continue in use, but extensions such as to 'nitroform' for trinitromethane are not desirable.

$-IO$ is iodosyl, $-IO_2$ iodyl, $-ClO$ chlorosyl, $-ClO_2$ chloryl, $-ClO_3$ perchloryl, and $-IX_2$ di-X-iodo.

X_2Hal^+ cations are named as substituted chloronium, bromonium, etc.

$COCl_2$ may be called phosgene, carbonyl chloride, or carbonic dichloride (*CA*), and $C(Hal)_4$ either a tetrahalomethane or a carbon tetrahalide.

Alcohols and Phenols

Alcohols and phenols are named by '-ol' suffixes if OH is the principal group, otherwise by hydroxy prefixes. Thus names such as 1,8-1,8-naphthalenediol and 8-quinolinol are more proper, as already noted on p. 72, than 1,8-dihydroxynaphthalene and 8-hydroxyquinoline, although both the latter still occur sometimes in the literature. *Chemical Abstracts* retains the trivial name phenol but forms all others systematically, e.g., 1-naphthalenol, 9-anthracenol, and 1,3-benzenediol; but abandonment of trivial names has not progressed so far in common usage.

In aliphatic alcohols the OH groups are cited as '-ol' when attached to any position in a main chain, e.g., 1-pentanol $CH_3(CH_2)_3CH_2OH$,

1,3-pentanediol $CH_3CH_2CH(OH)CH_2CH_2OH$. Presence of an OH group may determine choice of the main chain (cf. pp. 69–70).

Radicofunctional nomenclature is still used (although not in *CA* indexes) for simple compounds such as methyl alcohol and benzyl alcohol, notably in the aliphatic and aryl-aliphatic series. For $(CH_3)_3COH$ it is simpler to use *tert*-butyl alcohol than 2-methyl-2-propanol — note that *tert*-butanol is incorrect since there is no *tert*-butane to which a suffix '-ol' could be added. $(CH_3)_2CHOH$ is called isopropanol in industry but this too is incorrect as a systematic name.

In such long-known and populated series as alcohols and phenols it is natural that there is a large number of trivial names, far too many to be listed here. A few special points should, however, be noted.

Several alkoxybenzoic acids and alkoxybenzaldehydes have well-known trivial names, e.g., anisic acid $CH_3OC_6H_4CO_2H$, anisaldehyde $CH_3OC_6H_4CHO$, veratric acid $3,4\text{-}(CH_3O)_2C_6H_3CO_2H$, vanillin 7, and piperonal 8. To complete these series we require the alcohols

$ArCH_2OH$, where the radical $ArCH_2$ will be named anisyl, veratryl, etc.; this gives, correctly, names anisyl alcohol, etc., to the alcohols; yet when the aryl groups $CH_3OC_6H_4-$, $(CH_3O)_2C_6H_3-$, etc., occur as substituents they also have often been called anisyl, veratryl, etc., which has led to great confusion. It is wiser therefore to name these two types as $ArCH_2-$ alkoxybenzyl and $Ar-$ as alkoxyphenyl.

Trityl is official for the radical $(C_6H_5)_3C-$ as a prefix but not for use in radicofunctional nomenclature, e.g., not trityl alcohol but triphenylmethyl alcohol.

Use of carbinol as a name for alkyl substitution products of CH_3OH, exemplified by dipropylcarbinol for $(C_3H_7)_2CHOH$, is explicitly condemned by IUPAC (and so presumably are the derived carbinyl radical names) but has persisted, evidently because it is found useful.

Salts of alcohols and phenols may be named with the ending '-olate', as in sodium methanolate CH_3ONa, sodium phenolate $NaOC_6H_5$. Alternatively such names may be derived by citing the cation, the alkyl or aryl radical, and the word 'oxide' in that order, as in sodium cyclohexyl oxide, $C_6H_{11}ONa$. For salts of some common hydroxy compounds (C_1–C_4 alcohols and C_6H_5OH) contracted names are recommended and familiar: potassium *tert*-butoxide $(CH_3)_3COK$; $(C_6H_5O)_2Ca$, calcium phenoxide.

Ethers

For ethers, radicofunctional nomenclature is normal in simple cases, e.g., dimethyl ether, ethyl 1-naphthyl ether, 1-naphthyl propyl ether (note the alphabetical order of the two radicals).

Prefixes for ether substituents RO— are formed simply by adding '-oxy' to the name of the radical R, e.g., heptyloxy, heptenyloxy; but the '-yl' is elided for the $C_1 - C_4$ saturated alkyl groups, as in methoxy, and also in phenoxy and the generic name alkoxy.

In substitutive nomenclature, which is used by *CA*, there is no suffix ending for ether groups, which accordingly must be stated as prefixes. Even simple ethers are treated in this way, e.g., 1-ethoxyethane.

Alkoxy-prefix names are preferred over the ether names when one component is much larger than the other or when several ether groups are attached to one backbone, as in 3β-methoxy-5a-cholestane or 2,3,5-trimethoxyquinoline.

Ether names are convenient not only for simple cases and symmetrical ethers (e.g., dibutyl ether rather than 1-butoxybutane) but also for ethers of polyhydric alcohols or phenols that have well-known trivial names, e.g., glycerol 1,3-dimethyl ether and phloroglucinol trimethyl ether. In such cases the -ol ending of the alcohol or phenol is best retained.

The prefix for the —O— group is 'oxy-' and for —O—O— is 'dioxy-' ROOH are R hydroperoxides and ROOR are R peroxides, where R is the radical name.

The products of adding ozone at olefinic linkages have long been called ozonides, e.g., $C_4H_8O_3$, 2-butene ozonide.

Acids and their Derivatives

ACIDS: GENERAL

All organic acids have names ending in '-ic acid' or '-ous acid'. There are semitrivial names for many hundreds of acids — saturated, unsaturated, and aromatic, hydroxy- and amino-carboxylic acids, even for a few sulfur acids; some date back to the 17th century. Lists, far from complete, will be found in the IUPAC rules. Here we shall confine consideration to systematic nomenclature.

A few compounds that have chemically acidic groups other than CO_2H, SO_3H, etc., have long had acid-type names. Some of these, e.g., picric acid, styphnic acid, and ascorbic acid, are too convenient to replace; but others such as cresylic acid (as alternative to cresol) have been eliminated except as a name for crude commercial methylphenols, and the incorrect usage should not be extended. Contrariwise, the amino acids usually have names such as glycine, cysteine, and ornithine which emphasize the amino rather than the acid function. Since acid derivatives are usually named by modifying the -ic acid or -ous acid ending, these trivial names pose some problems.

Before dealing with acids and their derivatives in more detail it may be well to explain the historical reasons why the nomenclature of such well known classes of compound varies from case to case in what might be expected to form a single coherent series. The -ic acid names are derived from the French, e.g., 'acide acetique', where acetique is an adjective, rendered acetic in English. We can, if we wish, regard this acetic as a 'pseudoradical' name and the 'acid' as a functional class name. Salts and esters, however, have names such as sodium acetate and ethyl acetate, also relics of history, patterned on the ancient sodium chloride, ethyl chloride of Berzelius. Acyl radical names such as acetyl, benzoyl, and cyclohexanecarbonyl fall simply into modern usage and lead to acetyl chloride and the like as normal radicofunctional names. Amide names such as N-acetylpiperidine simply involve the radical acetyl as prefix, whilst acetamide and the like can be accepted as words coined to indicate that the compounds they represent are derived from the acids and ammonia. Finally, nitrile names are obviously derived from the 'nitr-' root, but the etymology of the '-ile' ending is not certain.

CARBOXYLIC ACIDS

All the usual and some unusual acids occurring in fats or oils, as well as the lower aliphatic acids and some simple aromatic acids, have semi-trivial names. From formic to valeric (also isovaleric) acid these are normally used throughout chemistry, but beyond C_5 systematic names are to be preferred for general use; in particular, the names caproic (C_6), caprylic (C_8), and capric (C_{10}) are so similar that they should be avoided.

Systematically, suffixes for CO_2H are formed in one of two ways: (1) the final '-e' of alkane, alkene, or alkyne, etc., is changed to '-oic acid', or (2) the CO_2H group is cited as a suffix '-carboxylic acid'.

(1)   In this nomenclature the C atom of the CO_2H group is considered as part of the original carbon chain; the ending '-oic acid' in fact

denotes a grouping **A** and not **B**. Thus nonane n-$C_8H_{17}CH_3$, for example, leads to nonanoic acid n-$C_8H_{17}CO_2H$, and the C of the CH_3 and the CO_2H receive number 1. For a,ω-dicarboxylic acids the ending -dioic acid is added to that of the hydrocarbon; nonane gives $HO_2C(CH_2)_7CO_2H$ nonanedioic acid (where no locants are needed).

(2)   The suffix '-carboxylic acid' denotes replacement of H by CO_2H, i.e., increase in the number of carbon atoms in the molecule. Thus n-$C_8H_{17}CO_2H$ might be, but rarely is, called 1-octanecarboxylic acid, and the CH_2 next to the CO_2H would then be numbered 1.

With these two types of name available, the general rules for principal groups explained on pp. 57, 59 and 69 would lead to the name 2-ethyl-pentanedioic acid (or 2-ethylglutaric acid) for **9**; but the carboxylic acid name for **10**, 1,2,4-butanetricarboxylic acid, neatly allows all the CO_2H groups to be cited in the same way. This is not possible for **11**, which has to be named 3-(carboxymethyl)hexanedioic acid.

$$CH_3CH_2\underset{\underset{CO_2H}{|}}{C}HCH_2CH_2CO_2H \qquad HO_2CCH_2CH_2\underset{\underset{CO_2H}{|}}{C}HCH_2CO_2H$$

9   **10**

$$HO_2CCH_2CH_2\underset{\underset{CH_2CO_2H}{|}}{C}HCH_2CO_2H$$

11

For CO_2H groups attached to cyclic structures the carboxylic acid nomenclature (or a semitrivial name) is obligatory, e.g., 1-pyrrolecarboxylic acid, 2,2'-biphenyldicarboxylic acid, as there is no CH_3 for conversion into CO_2H.

-oic/-ic

The reader will be familiar with many acid names ending simply in '-ic acid', not '-oic' or '-carboxylic'. These include all the trivial (semi-trivial) names of aliphatic acids but not their systematic equivalents. Among cyclic acids there is no regularity, e.g., benzoic, toluic, naphthoic, phthalic, cinnamic, furoic, nicotinic. It should be noted that *CA* has retained only four trivial acid names in its indexes, formic, carbonic, acetic, and benzoic; all other carboxylic acids are, except the biologically significant amino acids, named systematically.

OTHER ACIDS

Acids containing the group $-C(O)OOH$ are called peroxyacids. In individual names (except, by usage, for performic, peracetic, and perbenzoic acids) the prefix peroxy- is used before the trivial or -oic acid name of the acid or interpolated before -carboxylic, as the case may be, e.g., peroxybutyric acid $CH_3 CH_2 CH_2 C(O)OOH$, cyclohexaneperoxycarboxylic acid $C_6 H_{11} C(O)OOH$, and diperoxyphthalic acid o-$C_6 H_4 [C(O)OOH]_2$. *Chemical Abstracts*, however, forms such names by changing '-oic acid' to '-peroxoic acid', or '-carboxylic acid' to '-carboperoxoic acid', as in ethaneperoxoic acid $CH_3 C(O)OOH$, and cyclohexanecarboperoxoic acid.

The names of the sulfur acids, $RSO_3 H$ (sulfonic), $RSO_2 H$ (sulfinic), and $RSOH$ (sulfenic), have terminations that can be modified in the same ways as carboxylic acids for formation of derivatives. The R groups are cited by the name of the parent compound RH, giving $RSO_3 H$ benzenesulfonic acid, etc. (not the radical form phenylsulfonic).

For phosphorus acids *see* p. 170.

PREFIXES

Prefixes for the acid groups are carboxy $HO_2 C-$, sulfo $HO_3 S-$, sulfino $HO_2 S-$, and sulfeno $HOS-$.

The nomenclature of acyl groups $RC(O)-$ varies according to what R is, whether the name is to be used radicofunctionally (i.e., in two-part names) or substitutively, and whose rules are followed. According to IUPAC rules the names for acid radicals are derived from those of acids by changing '-oic acid' or '-ic acid' to '-oyl', or by changing '-carboxylic acid' to '-carbonyl'. For the trivial but usual names of the $C_1 - C_5$ acyclic acyl groups the ending is merely '-yl' instead of '-oyl', and '-yl' is also used instead of '-ine' to indicate modification of a trivially named amino acid to the acyl radical. This produces names such as hexanoyl, phthaloyl o-$C_6 H_4 (CO-)_2$, cyclohexanecarbonyl, acetyl, and alanyl, and complete names such as hexanoyl chloride and acetyl cation. The IUPAC substitutive prefixes have the same form except when derived from '-carboxylic acid', when $RC(O)-$ is named with a compound prefix in which R gives the radical name and CO is cited as 'carbonyl' (as usual); thus one arrives at, for example, 1-(cyclohexylcarbonyl)piperidine.

The *CA* radicofunctional names differ in nearly all cases from those specified by IUPAC. In *CA* indexes only three such names — formyl, acetyl, and benzoyl — are trivially based. Acyl groups derived from acids named by the 'carboxylic acid' system are called cycloalkylcarbonyl and arylcarbonyl groups (not cycloalkanecarbonyl and arenecarbonyl as in IUPAC rules), e.g., fluorenylcarbonyl. All other RC(O)– groups regarded as substituents are named as 1-oxoalkyl groups, e.g., 1-oxopropyl $CH_3 CH_2 C(O)–$ (formerly propionyl or propanoyl) and 1,4-dioxo-1,4-butanediyl $–C(O)CH_2 CH_2 C(O)–$ (formerly succinyl or butanedioyl). These 1-oxoalkyl names describe structure but, being unfamiliar, may tend to obscure the functional relationship to acids.

The distinction between radicofunctional names and substitutive names for acid radicals also applies to the sulfur acids; for the compound prefixes for $X–SO_2 –$, $X–SO–$, and $X–S–$, sulfonyl is used for $>SO_2$, sulfinyl for $>SO$, and thio for $>S$. Thus $C_6 H_5 SO_2 –$, for example, is phenylsulfonyl when operating as a substituent.

Similar compound prefixes are used in many other cases. Very important is the ester prefix $RO_2 C–$, clearly named as alkoxycarbonyl-, e.g., methoxycarbonyl-, benzyloxycarbonyl-, but $ClOC–$ is chloroformyl. Since $NH_2 CO_2 H$ is carbamic acid, $NH_2 C(O)–$ has the IUPAC name carbamoyl.

IONS

Anions RCO_2^- and the analogous sulfur anions are named by changing '-ic acid' to '-ate', giving an ending '-ate' or '-oate' according to whether the acid name ends in '-ic' or '-oic'. Examples are sodium acetate $CH_3 CO_2 Na$, potassium benzenesulfonate $C_6 H_5 SO_3 K$, potassium decanoate, and triammonium 1,2,4-benzenetricarboxylate.

Salts of amino acids having trivial names are best named by periphrasis such as glycine sodium salt, although names of the type sodium glycinate are common and not so obviously wrong in the light of modern inorganic nomenclature.

The prefix for the ion $^-O_2 C–$ is carboxylato.

ESTERS

By an analogy with salts that has been known to be wrong for close to 100 years, esters are named analogously to salts, the radical name for the R of $RO–C(O)–$ replacing the name of the cation of a salt, as in methyl acetate.

Prefixes for ester groups were mentioned above. The other kind of prefixes for such groups, $RC(O)O–$, are named merely by adding '-oxy' to whatever acyl group name is appropriate, e.g., $C_6 H_5 C(O)O–$, benzoyloxy; $C_6 H_{11} C(O)O–$, cyclohexylcarbonyloxy.

AMIDES

For compounds $RCONH_2$, RSO_2NH_2, etc., the ending '-ic', '-oic' or '-ylic' acid is changed to '-amide', giving acetamide, hexanamide, 1-pyrrolecarboxamide, benzenesulfonamide, and so on. When the N is substituted, giving $RCONHR'$, $RSO_2NR'_2$, etc., the compound may be named as an *N*-substituted amide, e.g., *N*-methylacetamide, *N,N*-diethylbenzamide. There is an alternative of citing the RCO group as an acyl radical, and this is done when the R' group on the nitrogen is more complex than the R group of the acyl; a simple case is 1-acetylpiperidine. In many cases choice between the two methods is optional.

When R' is a phenyl group, names such as acetanilide and sulfanilanilide may be used.

Monoamides of common dicarboxylic acids may be given names ending in '-amic acid' or '-anilic acid', e.g., phthalamic acid o-$NH_2COC_6H_4CO_2H$, succinanilic acid $C_6H_5NHCOCH_2CH_2CO_2H$; but *CA* (wisely) uses the longer, more systematic names 2-(aminocarbonyl)benzoic acid and 4-oxo-4-(phenylaminocarbonyl)-butanoic acid, respectively.

Various situations of further complexity arise or can be thought up for acyl derivatives of amines, and some of these are discussed in the IUPAC rules[1c].

DERIVATIVES OF DIBASIC ACIDS

In names of derivatives of dibasic acids it is customarily assumed, sometimes unjustifiably, that both acid groups are involved. Thus sodium succinate denotes the disodium salt, malononitrile is $CH_2(CN)_2$, and phthalamide is the diamide. Monometal salts are treated as acid salts, e.g., sodium hydrogen succinate, and esters similarly; but for many types of derivative one group must be cited as prefix, as in p-(chloroformyl)benzoic acid $ClCOC_6H_4CO_2H$.

Lactones, Lactams, and Their Sulfur Analogues

There are many ways of naming these compounds that have caused much trouble, so here we shall outline only the methods that have wide use and IUPAC approval.

Intramolecular cyclic esters, in which the hydroxyl group and the carboxyl group of a hydroxy acid are considered to have interacted with loss of water, are called lactones.

Lactones derivable from monohydroxy acids may be named as lactones or by an '-olide' ending; compound **12** may be called δ-valerolactone or 5-pentanolide.

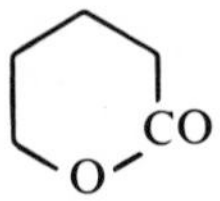

12

More complex, particularly polycyclic, compounds may be named as carbolactones, this suffix denoting introduction of a −CO−O− group with the CO taking the lower locant; for example, compound **13** would be 4,3-pyrenecarbolactone.

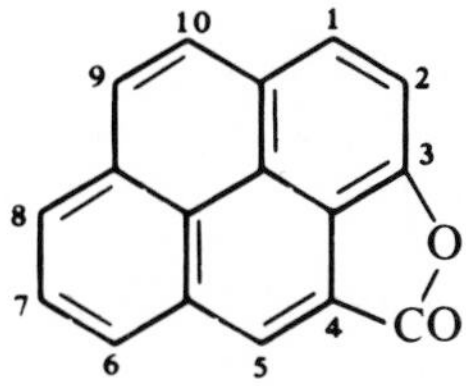

13 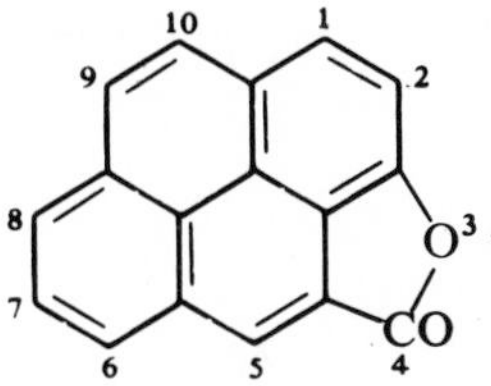   **13a** 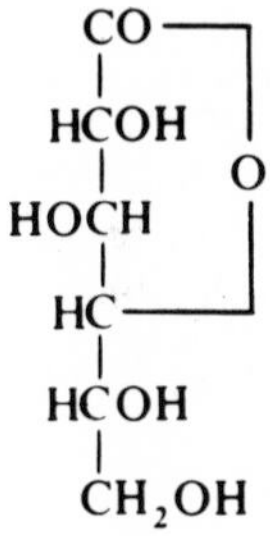   **14**

When a polyhydroxy acid has a trivial name, a derived lactone name may be patterned on D-glucono-1,4-lactone for **14**.

Compounds that have a group −CO−NH− or −C(OH)=N− as part of a ring are preferably named as heterocyclic compounds but may be named lactones but with 'lactam' or 'lactim', respectively, as suffix.

Sultones are internal esters in which $>SO_2$ replaces the $>CO$ of a lactone and may be named in the same way as lactones but with '-sultone' replacing '-carbolactone' as suffix to the parent name.

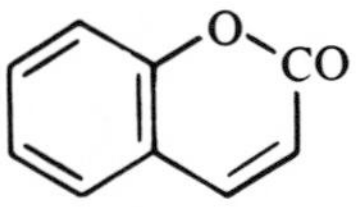

15 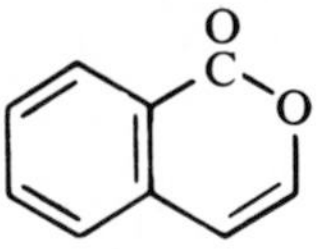   **16** 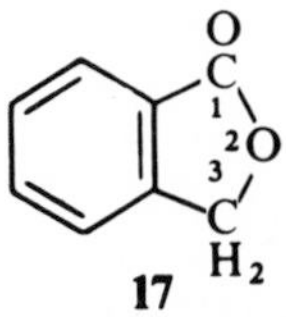   **17**

Sultams are the SO_2 analogues of lactams and have the suffix '-sultam' replacing '-lactam'. It is undesirable to extend this system to other cyclic esters and amides, e.g., cyclic sulfites should not be called sultines.

All these compounds may, of course, be also named as heterocycles by the ordinary methods, e.g., **12** as tetrahydro-2-pyrone (*CA* name, tetrahydro-2*H*-pyran-2-one), and **13** (renumbered as **13a**) as 4*H*-pyreno[3,4-*bc*]furan-4-one.

Many lactones have trivial names, notably coumarin **15**, isocoumarin **16**, and phthalide **17**.

Nitriles and Derived Groups

By one method, confined to the aliphatic series, nitriles can be named by adding '-nitrile', to signify $\equiv N$, to the name of the hydrocarbon with the same number of carbon atoms; this is analogous to the '-oic acid' nomenclature and gives, for example, nonanenitrile $n\text{-}C_8H_{17}CN$ and nonanedinitrile $NC(CH_2)_7CN$.

The carboxylic acid nomenclature is mirrored in the ending '-carbonitrile' for the $-CN$ group, as in 1-pyrrolecarbonitrile C_4H_4N-CN and 1,2,4-butanetricarbonitrile (analogous to the acid **10**).

By a third method, when the corresponding acid has a trivial name, the ending '-ic acid' or '-oic acid' is changed to '-onitrile', as in acetonitrile and benzonitrile.

The prefix for a NC— group is 'cyano-'.

Radicofunctional names in which cyanide denotes the $-CN$ function are normally used only in special and simple cases such as benzoyl cyanide C_6H_5COCN. According to IUPAC rules, when compounds contain the other groups in *Table 6.2* as principal group they are named radicofunctionally, e.g., phenyl isocyanate C_6H_5NCO, but otherwise the group is cited as prefix, as in thiocyanatoacetic acid,

Table 6.2 CYANIDE AND RELATED GROUPS IN ORDER OF DECREASING PRIORITY FOR CITATION AS FUNCTIONAL CLASS NAME

Group X in RX	Functional class ending and generic name of class	Prefix
—CN	cyanide	cyano
—NC	isocyanide*	isocyano
—OCN	cyanate	cyanato
—NCO	isocyanate	isocyanato
—ONC	fulminate	—
—SCN	thiocyanate	thiocyanato
—NCS	isothiocyanate	isothiocyanato
—SeCN	selenocyanate	selenocyanato
—NCSe	isoselenocyanate	isoselenocyanato

*Not isonitrile or carbylamine.

$NCS-CH_2CO_2H$. *Chemical Abstracts*, however, names all these groups (except $-CN$) exclusively with prefixes, e.g., C_6H_5NCO, isocyanatobenzene.

Aldehydes

Aldehydes are named by one of the following methods:
(1)   by a suffix '-al' to the name of an aliphatic hydrocarbon (with the usual elision of e) to denote a grouping **C**,

(2)   by adding '-carbaldehyde' to the name of a hydrocarbon (aliphatic or cyclic) or heterocycle to denote the group $-CHO$, or
(3)   by changing the ending '-oic acid' or '-ic acid' of a semitrivial acid name to 'aldehyde'.

The procedures (1) and (2) are as for acids and are used by *CA* * except for the three commonest members of the family formaldehyde, acetaldehyde, and benzaldehyde; other such common names, often seen, are of course generated by procedure (3).

A few aldehydes have specific trivial (e.g., vanillin) or semitrivial (e.g., piperonal) names.

Oximes, semicarbazones, hydrazones, and similar derivatives of aldehydes are named by citing the name of the derivative as a separate word, as in nonanal semicarbazone and benzaldehyde 2,4-dinitrophenyl-hydrazone. Occasional abbreviations such as acetaldoxime and benzaldoxime are still met but seem unnecessary, and such names for analogues should not be proliferated.

Some compounds containing both aldehyde and carboxyl groups have names of the type illustrated by succinaldehydic acid (cf. -amic acids, p. 112).

'Formyl-' is the prefix name for $OHC-$, this being also the acyl radical of formic acid.

Ketones

In substitutive names of ketones, the $=O$ group is cited as suffix '-one' or prefix 'oxo-' according to whether it does or does not signify the principal group; examples are 2-butanone $CH_3CH_2COCH_3$ or

*Who, however, have recently changed to '-carboxaldehyde'.

4-oxocyclohexanecarboxylic acid. *Chemical Abstracts* uses such
substitutive nomenclature. For a compound such as $CH_3COC_6H_5$,
this produces the name 1-phenyl-1-ethanone, unusual because such
1-ethanone parents when unsubstituted are of course called aldehydes.
The older prefix 'keto-' is obsolescent but persists in generic names
such as keto acids.

In radicofunctional nomenclature the $>C=O$ group is the function,
and the class name is ketone, which is preceded by separate words
specifying the radicals R and R' of $RR'CO$; thus $CH_3CH_2COCH_3$
becomes ethyl methyl ketone (note the alphabetical order, ethyl
methyl).

Trivial and semitrivial names are not quite so common as in some
other series, but several are important; acetone CH_3COCH_3, benzil
$C_6H_5COCOC_6H_5$, deoxybenzoin **18**, chalcone **19**, and ketene
$CH_2=CO$ (not functionally a ketone) spring to mind (note the locants
in **18** and **19**). Benzil has been made the pattern also for others, such
as furil $[(2-C_4H_3O)CO]_2$ where C_4H_3O is furyl, and there is benzoin
$C_6H_5CH(OH)COC_6H_5$ and its analogues.

18   **19**

So far so simple, but there are complications to follow.

First, it will be obvious that still another method of naming is
possible; $RCO-$, being an acyl group, may be named as such when
substituted into the compound represented by R', and this is common
when R' is large and cyclic and the acyl group relatively small. Indeed,
as an example, 2-acetylpyrene seems the best name for the ketone **20**.
There is a long-standing variant of this, whereby for benzene and
naphthalene derivatives the '-yl' of a small acyl group is changed to 'o',
and '-phenone' and '-naphthone' are used for the ring system into which
the acyl group is substituted; as familiar examples we may cite
acetophenone $C_6H_5COCH_3$, benzophenone $C_6H_5COC_6H_5$, and
2'-propionaphthone for **21** [note the priming (dashes) of the ring
locants].

It is, however, cyclic ketones that require the most care. To start
with, a CO group next to O or N characterizes, respectively, a lactone
or a lactam and such compounds can be named as belonging to those
classes; but in fact, many lactones and most lactams are usually named
as if they were ketones. Thus the lactam **22** is called 2-piperidone, but
the lactone **12** is ordinarily named δ-valerolactone, as noted above.

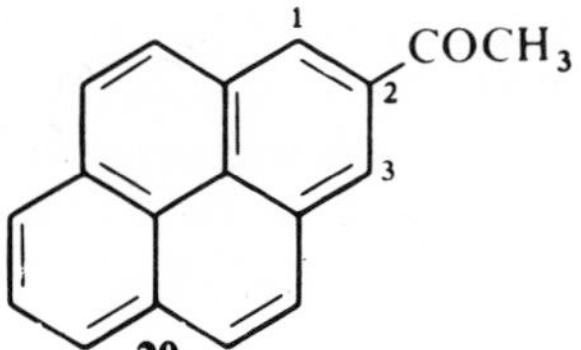

20

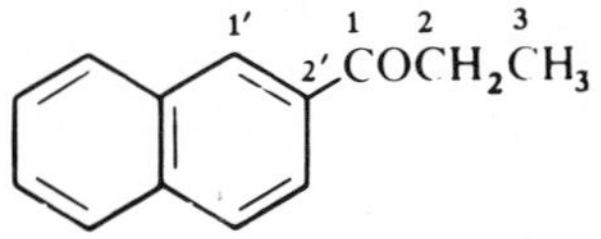

21

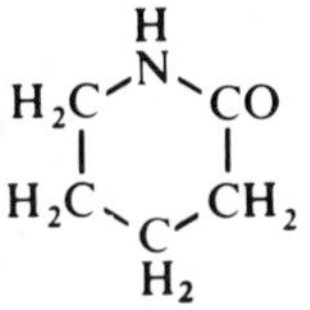

22

Chemical Abstracts indexes **22** at 2-piperidinone and **12** as tetrahydro-2*H*-pyran-3-one, as already noted.

It was explained in Chapter 5 that x*H* (where x is a locant and *H* is termed indicated hydrogen) is placed before the name of a cyclic compound when an extra hydrogen atom is required after the maximum number of conjugated double bonds has been inserted into the formula of a cyclic structure. Such situations arise, for example, with indene **23** and carbazole **24**. Both these compounds are so common that the 1*H* of indene **23** and the 9*H* of carbazole **24** are normally omitted from the names; however, the isomers **23a** and **24a** would be named 4*H*-indene and 1*H*-carbazole, respectively, and would cease to be hypothetical if the H_2 were replaced by, say, two methyl groups.

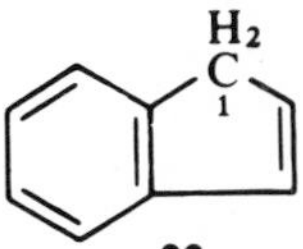

23

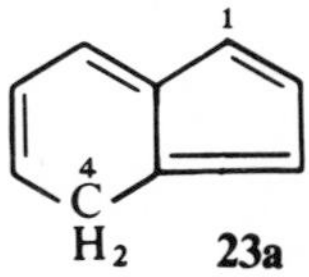

23a

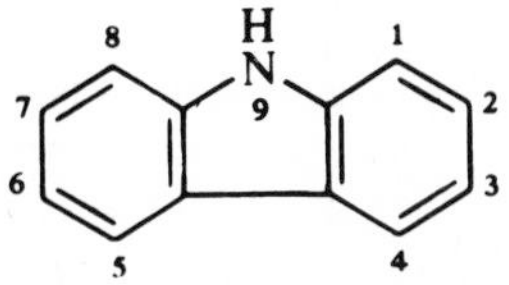

24

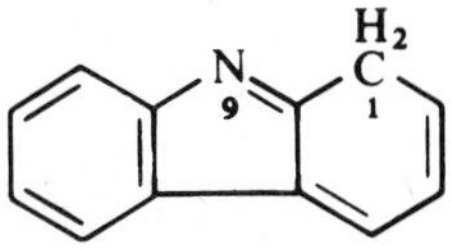

24a

Now since a suffix -one denotes replacement of 2H by =O, the ketone **25** becomes 1-indenone, and **26** becomes 4-indenone. Here the indicated hydrogen prefix may be regarded as unnecessary in view of the locant of the CO group; but *CA* retains the name of the hydrocarbon or heterocycle, with hydrogen indicated, and then adds the suffix: thus 1*H*-inden-1-one.

A different trouble meets us if we introduce an oxo group into a parent ring system that does *not* require extra hydrogen, say, pyrene, giving, say, **27** or **28**, and the position of this hydrogen must be indicated in the name, by means of *H*. But, as this hydrogen arises owing to introduction of the oxo group and is not due to the pyrene skeleton itself, the *H* in the name should be associated with the '-one' suffix and not with the name pyrene. This is done as in 2(1*H*)-pyrenone **27** and 2(7*H*)-pyrenone **28**.

Similar situations arise, of course, with heterocyclic ketones. The ketone **29** requires the name 2*H*-carbazol-2-one, but the ketone **30** must be named 3(2*H*)-phenanthridinone.

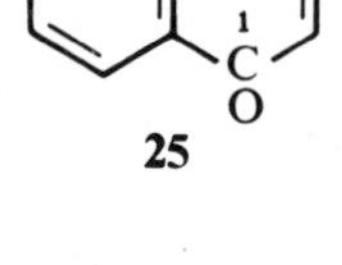

25

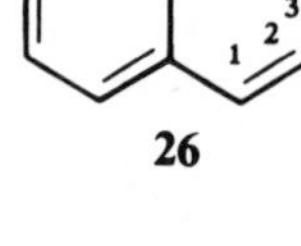

26

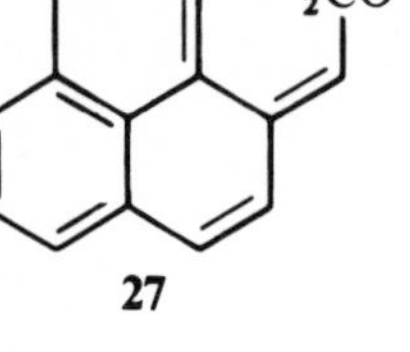

27

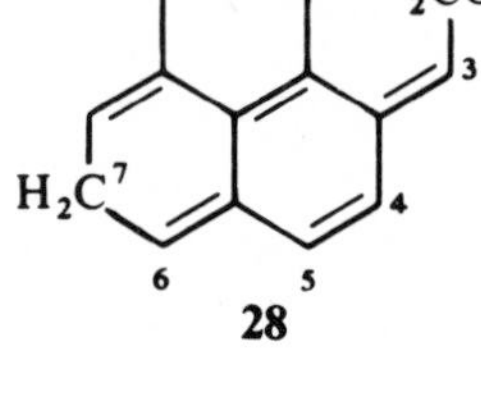

28

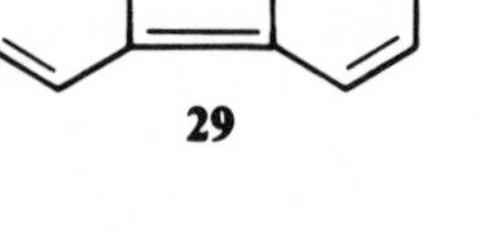

29

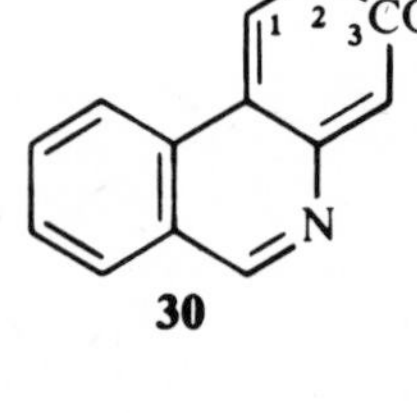

30

A small digression must be made before proceeding further. Some ketone names may be abbreviated by IUPAC rules (but not by *CA*); common cases are anthrone (for 9(10*H*)-anthracenone) and omission of 'in' from acridone, pyridone, piperidone, quinolone, and isoquinolone; in some cases the indicated hydrogen is also omitted. The long-established 4- **31** and 5-pyrazolone **32** do not conform to the rules since two H atoms can be removed from each ring, but in spite of this these names are widely used. The simplified names 4-oxazolone, 4-isoxazolone, and 4-thiazolone signify compounds with $CO-CH_2-O$ or $CO-CH_2-S$ sequences in the ring, and thus these compounds do not contain maximum conjugated unsaturation.

31   **32**

The nomenclature of certain cyclic di- and poly-ketones presents some special features. Aromatic unsaturated diketones or tetraketones containing complete conjugation have long been named by adding the suffix '-quinone' to the name of the parent hydrocarbon or heterocycle (the parent name sometimes shortened as in anthraquinone, **33**). *Chemical Abstracts* as usual names **33** systematically as 9,10-anthracenedione.

More serious trouble to authors can arise among, e.g., pyridine derivatives. Compound **34** is 2(1*H*)-pyridone, though the 1*H* is usually omitted; no extra hydrogen is needed in **35**, which is simply 2,5-pyridinedione. But **36** (also a diketone) is 2,4(1*H*,3*H*)-pyridinedione and **37** could be 2,4,6(1*H*,3*H*,5*H*)-pyridinetrione; but since the saturated heterocycle has the name piperidine it is simpler to call **37** 2,4,6-piperidinetrione. Note also that **38** becomes 3,4-dihydro-2(1*H*)-pyridone.

These names illustrate why the nomenclature of cyclic ketones has been described as 'a semantic disaster area'[2]. The anomalies are obvious; the series **34–36** are all named as pyridine derivatives, although **34** and **35** are derivable from dihydropyridines and **36** is derivable from a tetrahydropyridine; **38** receives a dihydro prefix although it is derivable from the same tetrahydropyridine. The state of hydrogenation of the parent ring system is not considered from a chemical viewpoint in this nomenclature, nor in the *CA* name for **33** − it is solely a matter of whether extra hydrogen is needed after first the oxo group and then the maximum number of non cumulative double bonds have been introduced into the ring.

Now, in the past, nomenclature has been used which did take regard
to the chemical state of hydrogenation of the ring; in it 33 was called
9,10-dihydro-9,10-dioxoanthracene, and 34 1,2-dihydro-2-oxopyridine.
But this hydro-oxo system has a fundamental failing, namely that it
does not provide a suffix or functional class name for the ketone group

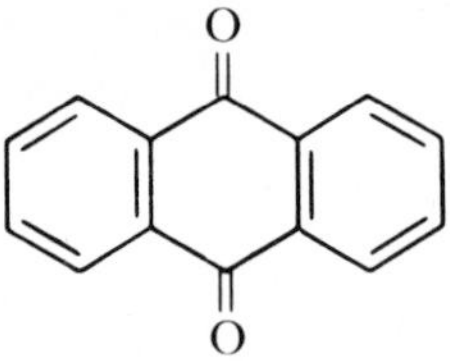

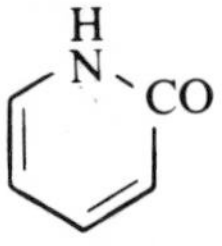

34

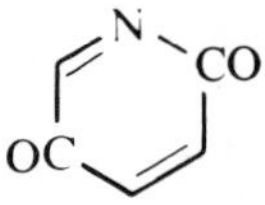

35

33

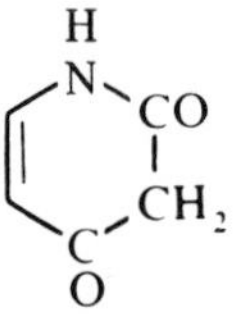

36

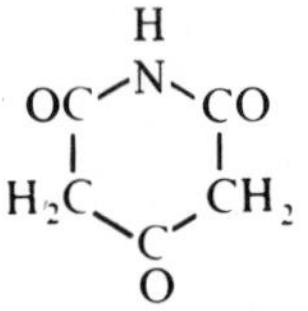

37

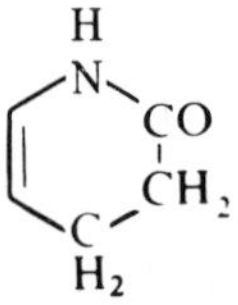

38

that is the principal group in these cyclic ketones. The hydro-oxo names
thus have no place in modern nomenclature, and in spite of the apparent
chemical anomalies the system outlined above, with H where necessary,
should be used today.

DERIVATIVES

Oximes, hydrazones, etc., are named as for aldehydes (*see* p. 115).

Amines

Amines have names ending in '-amine'. This ending is best used as a
suffix in substitutive nomenclature, like '-ol' or '-one', to modify the
name of a parent hydrocarbon or heterocycle:

ethanamine	$CH_3CH_2NH_2$
1,3,5-pentanetriamine	$H_2NCH_2CH_2CH(NH_2)CH_2CH_2NH_2$
2-pyridinamine	$2\text{-}H_2NC_5H_4N$
benzenamine	$C_6H_5NH_2$

(These names are not yet as familiar as the ones also authorized by IUPAC, based on Hofmann's contribution long ago to the theory of types, in which amines are regarded as substitution products of ammonia: ethylamine, ethylmethylamine, trimethylamine, N-methylbenzylamine. Nevertheless, it is better to get rid of this old irregularity and use carbon compounds as parents.) The suffix system is now used exclusively by *CA*, although the name aniline will not soon be displaced from common usage. For secondary and tertiary amines, the largest hydrocarbon parent is chosen (if there is a choice), and other groups on nitrogen are cited as prefixes, with N or N,N- as locant(s):

$CH_3 CH_2 NHCH_3$ N-methylethanamine
$C_6 H_5 N(C_2 H_5)_2$ N,N-diethylbenzenamine
$(CH_3)_3 N$ N,N-dimethylmethanamine

The names of compounds containing nitrogen as part of a cyclic skeleton end in one of the syllables listed for them in *Table 5.6*. (In heterocyclic compounds containing an NH group, the numerical locant is preferred to N-, as in 1-phenylimidazole.)

The prefix for $H_2 N-$ in nomenclature is 'amino-', but the free radical $H_2 N\cdot$ is named aminyl (*see* p. 102).

There are many semitrivial names for amines, notably for alkaloids, other heterocyclic bases, and amino acids. A number of substituted benzenamines have long been named with the suffix '-idine', as in xylidine and anisidine, but most special names for such compounds are falling into disuse, except toluidine, benzidine (which is not simply derived from benzene), and of course aniline.

AMINE SALTS

The IUPAC Organic rules[1d] do not plainly specify a general system for naming salts of primary, secondary, and tertiary amines, but the Inorganic rules[3] prescribe naming them as substituted ammonium salts, e.g., $[(C_2 H_5)_2 NH_2]^+ Cl^-$, diethylammonium chloride. Such names are unambiguous and sometimes seen. However, it seems best to follow *CA* usage here and to name ions formed by protonating primary, secondary, and tertiary amines by changing the final '-e' of the amine name to 'ium' (cf. p. 103). This yields names such as methanaminium chloride for $[CH_3 NH_3]^+ Cl^-$ and pyridinium nitrate for $[C_5 H_5 NH]^+ NO_3^-$. Obviously the name methylaminium chloride is also intelligible. Care must be taken in naming amine salts of dibasic acids; for example, $[C_6 H_5 NH_3]^+ HSO_4^-$ is anilinium hydrogen sulfate

and $2[C_6H_5NH_3]^+_. SO_4^{2-}$ is dianilinium sulfate, whereas aniline
sulfate or even anilinium sulfate might refer to either salt.

Occasional use is still made of names such as aniline hydrochloride,
pyridine picrate (note that hydro- is included only with the halide
names; aniline hydrochloride, hydrobromide, or hydriodide, but
aniline picrate or nitrate; this old, irrational practice is probably
derived from the names of the acids).

Quaternary salts are named by IUPAC rules as ammonium salts, e.g.,
$[(CH_3)_4 N]^+ Br^-$, tetramethylammonium bromide, but *CA* now con-
siders them derivatives of the 'senior' primary amine present, so that
the salt just mentioned must be called *N,N,N*-trimethylmethanaminium
bromide, $[C_6 H_5 N(CH_3)_3]^+$ the *N,N,N*-trimethylbenzenaminium ion,
etc. Quaternary salts are sometimes handled by an older method, as
metho-salts, e.g., triethylamine methochloride, quinoline methopicrate;
such obsolescent names should not normally be used, though they are
useful for compounds of uncertain structure.

Sulfur Compounds

Sulfonic, sulfinic, and sulfenic acids have been mentioned on p. 110,
but there are many further aspects of organic sulfur compounds for
which IUPAC nomenclature is complex, often little understood, and
generally unsatisfactory. Consider the following bare facts.

The syllables 'thio', to denote sulfur, appear in thiol, thione,
thionia, and as thio itself; and 'thia-', as already mentioned, denotes a
sulfur atom in a ring on the Hantzsch–Widman system (*see* p. 95) or
replacement of carbon by sulfur in replacement nomenclature (*see*
pp. 56, 98).

'Thiol', as a suffix, denotes –SH present as principal group in
substitutive nomenclature (*Table 4.3*), as in methanethiol $CH_3 SH$;
but 'mercapto' is used for HS– as prefix.

Thione can be used to denote =S doubly bonded to one atom when
it is the principal group, just as -one denotes carbonyl =O, e.g.,
2-butanethione for $CH_3CH_2 C(=S)CH_3$. Nevertheless, the generic name
for such compounds is thioketones and the prefix for this =S is
'thioxo-'!

Sulfonium denotes a cation $[\lambda^3 -S]^+$, as in trimethylsulfonium
chloride $[(CH_3)_3 S]^+ Cl^-$. Sulfonio- is used when $R_2 S^+-$ is named as
a prefix. 'Thionia-' is the prefix for a λ^3-S cation in replacement nomenclature (*see* p. 172).

Thio- (unmodified) requires more consideration. It is used in two
distinct ways: to denote –S– as a linear component and to replace
oxygen. (It is the latter that leads to 'thioketone' and 'thioxo-'.)

Linking the bivalent thio $-S-$ to one organic group gives the radical RS$-$, which is named 'R-thio', e.g., methylthio- for CH_3S-, in the same way as the bivalent radical $>CO$ carbonyl leads to methoxycarbonyl for CH_3O_2C-. Radical names of 'methylthio-' type are placed as prefix when this group is attached to a large group, as in 1-(methylthio)pyrene. There is naturally a radicofunctional alternative to this, namely the use of sulfide as the functional class name (cf. inorganic chemistry which led to this older nomenclature), preceded by the names of radicals of RR'S as separate words, as in ethyl methyl sulfide $CH_3CH_2SCH_3$ or dimethyl sulfide $(CH_3)_2S$; such names are still used for simple compounds.

Salts of thiols are thiolates, e.g., CH_3CH_2SNa sodium ethanethiolate or, by radicofunctional nomenclature, sodium ethyl sulfide.

It will then be apparent that, if AlkS$-$ is 'alkylthio-' then an HS$-$ group could have been called 'hydrothio'; but this was rejected in favour of 'mercapto-'; yet 'mercapto-' was derived from mercaptan for RSH, an old name that was itself rejected in favour of thiol!

A second type of use of 'thio-' for $-S-$ is in naming symmetrical compounds; thus $S(CH_2CH_2CO_2H)_2$ is simply named as 3,3'-thio-dipropionic acid.

More diverse, and sometimes unfortunate, are the applications of thio to denote replacement of oxygen, in effect for naming sulfur analogues of oxygen compounds. Most of the difficulties arise when thio is used to indicate replacement of oxygen singly bonded to two groups or atoms. Thus thiophenol C_6H_5SH looks like the systematic name for a hydroxy derivative of thiophene. Again, compound **39** is 4H-pyran, so with 'thio' replacing $-O-$, compound **40** becomes 4H-thiopyran; but tetrahydro-4-thiopyranone might mean either **41** or **42**. While **41** is logically called tetrahydro-4-pyranthione, **42** is preferably given the Extended Hantzsch–Widman name 4-thianone,

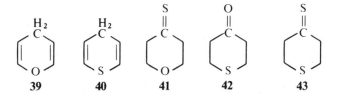

39   **40**   **41**   **42**   **43**

and similarly **43** is 4-thianethione. It may be noted that a different solution, namely thiapyran for **40**, had been used earlier, but this was rejected by IUPAC as opening the door too wide to replacement of other heteroatoms by sulfur, which might, e.g., lead to 4H-thiapyridine as an alternative name for **40**.

A special case is provided by thioglycolic acid, a name sometimes applied to $HSCH_2CO_2H$ as the sulfur analogue of glycolic acid $HOCH_2CO_2H$. The difficulty is that thioglycolic acid is also the name (*see* below) of a glycolic acid in which an oxygen atom of the $-CO_2H$ group has been replaced by sulfur. Here the solution is to use the unambiguous name mercaptoacetic acid for $HSCH_2CO_2H$.

$$CH_3C\left\{{O \atop S}\right\}H^{\,*}$$

D

There are analogues for both these special cases.

'Thio' and 'dithio' are also widely used in names of thio acids, both inorganic (*see* p. 19) and organic, such as thioacetic acid **D** and dithiobenzoic acid $C_6H_5CS_2H$ to denote replacement of one or both of the oxygen atoms of the CO_2H group. The dithio name is unambiguous; the monothio name does not specify whether the $=O$ is converted into $=S$ or the OH into SH, and precisely this indecision is intended because it is normally not known whether the H is attached to O or S in the thio acid (though presumably more to S than to O). If this distinction is known, the species are to be called, for example, thioacetic *S*-acid for $CH_3C(=O)SH$ and thioacetic *O*-acid for $CH_3C(=S)OH$. The distinction is, however, usually known for esters, and here it is easily indicated by such names as *S*-methyl thioacetate $CH_3C(=O)SCH_3$ and *O*-methyl thioacetate $CH_3C(=S)OCH_3$. For acids named with the '-carboxylic acid' suffix the sulfur analogues take the endings '-carbothioic acid' and '-carbodithioic acid'. Previous distinction by means of the endings '-thiolic', '-thionic', and '-thionothiolic' has been discarded by IUPAC.

The names R-sulfonyl and R-sulfinyl for the radicals RSO_2- and $RSO-$ (R is the radical name), respectively, have been mentioned on p. 111. These should be contrasted with sulfone as the radicofunctional class name for an $>SO_2$ group, as in $CH_3CH_2S(O_2)CH_3$ ethyl methyl sulfone, and sulfoxide correspondingly for an $>SO$ group.

As prefix names for RS_2-, RS_3-, etc., there are the names 'R-dithio-', 'R-trithio-', etc. (R is the radical name); disulfide, trisulfide, etc., then replace sulfide in radicofunctional names such as ethyl methyl disulfide $CH_3CH_2S_2CH_3$. Recently a new nomenclature has been developed for compounds known or believed to contain chains of sulfur atoms, based on regarding HSH as sulfane. HSSH is termed disulfane, HSSSH trisulfane, and so on (cf. p. 14). Groups or atoms

*Structure **D** is intended to show by formula, as thioacetic acid does by name, that the hydrogen atom may be linked to either the oxygen or the sulfur atom.

replacing the hydrogen are cited as prefixes. Thus C_6H_5SSH is phenyldisulfane and $CH_3CH_2S-S-SCl$ can be called chloroethyltri-sulfane; a prefix is used even if the compound is acidic, as for CH_3SSSOH, which is called 1-hydroxy-3-methyltrisulfane; the salt CH_3SSSNa is sodium methyltrisulfanide and a salt $CH_3SSSONa$ is sodium 3-methyl-1-oxidotrisulfane ('oxido-' is $-O^-$).

Indicated Hydrogen — Generalization of the Concept

Indicated hydrogen, symbolized by H preceded by a locant, has been discussed in the pages above in two connections. One (*see* pp. 82, 91) was in names such as 1*H*-fluorene or 4*H*-pyran, where after introduction of as many non-cumulative double bonds as possible into a ring system, there remains 'extra' hydrogen to be introduced and there is more than one possible position for that hydrogen, so that its actual position must be 'indicated' by a locant. The other occasion (*see* pp. 117–119) was with ketones; here one met ketones formally derived by replacement of CH_2 by CO in a ring system already carrying an indicated hydrogen, so that the oxo group is named merely by a suffix -one, as in 1*H*-fluoren-1-one; but cases were also found where indicated hydrogen was required after, but not before, the ketonic oxygen group was placed on the ring skeleton; the latter case is exemplified by naphthalene **44** and 2(1*H*)-naphthalenone **45**, where the symbolism 2(1*H*) shows that the 1*H*-hydrogen is present because of the 2-oxo group.

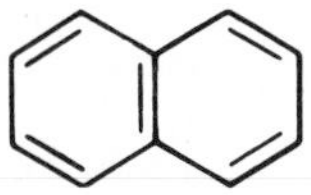

44 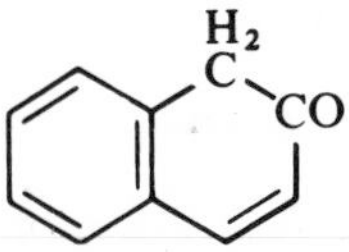   **45**

Though simple cases such as those discussed above and previously are the most often encountered, complications and extension of these procedures are not infrequent.

Complications of the ring structure alone may be noted: the spiran **46** which is named spiro[cyclopentane-1,2'-2*H*-indene] (so as to pin down the isomeric indene skeleton); also the radical **47** which is

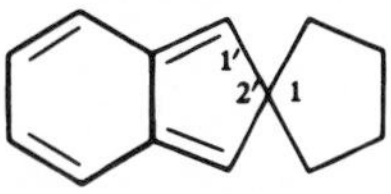

46 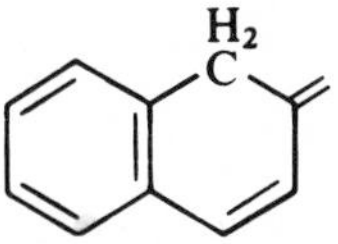   **47**

2(1*H*)-naphthylidene, the symbolism here being 2(1*H*) to show that
the need for the extra 1*H*-hydrogen arises from introduction of the
-ylidene radical.

Compound **48** is analogous to the simple ketones and is named as a
derivative of 2*H*-carbazole — simply 2*H*-carbazole-2,2-dicarboxylic
acid. The general case corresponding to the ketone **45** arises when *any*
principal group is introduced where there was no hydrogen in the
parent compound, or only insufficient hydrogen. For example, **49** is
2,2(1*H*)-naphthalenedicarboxylic acid, the isoquinoline derivative **50**
is 2(1*H*)-isoquinolinecarboxylic acid, and **51** is 8a(1*H*)-isoquinoline-
carbonitrile.

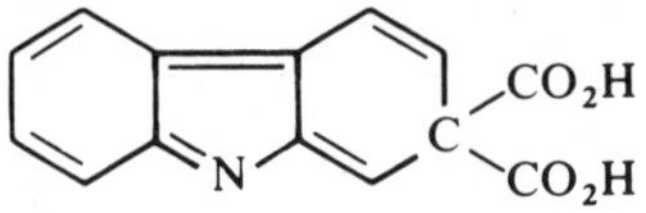

48

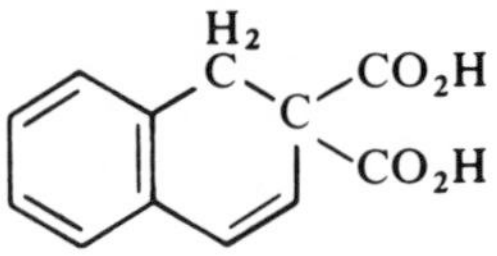

49

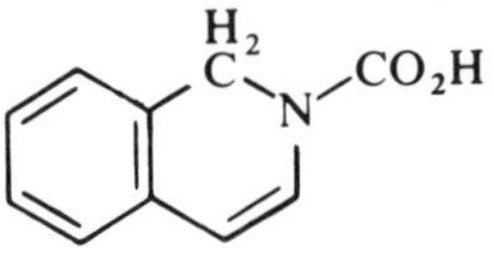

50

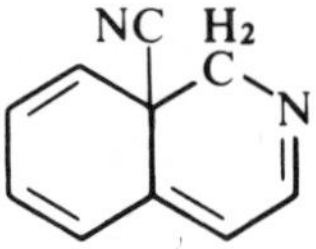

51

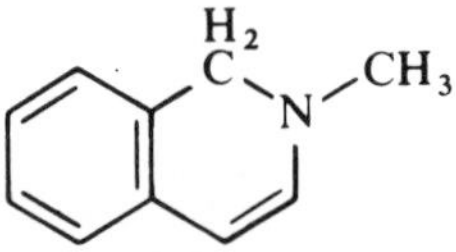

52

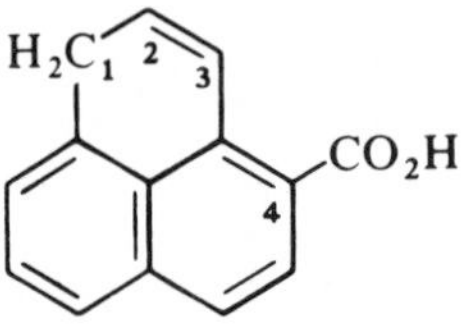

53

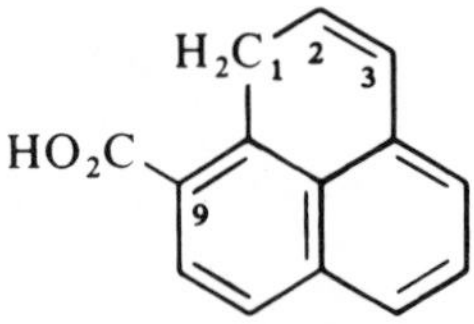

54

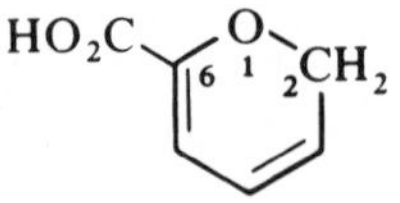

55

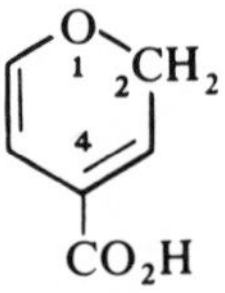

56

Perhaps the most perplexing feature is that this 2(1H) type of procedure is applied only in the presence of principal groups, i.e., when the introduced group is named as a suffix. The compound **52** is called 1,2-dihydro-2-methylisoquinoline, in contrast to **50**.

Finally we may note that indicated hydrogen has the highest priority for lowest locant (cf. p. 64) after any fixed numbering, higher even than the position of a radical valence or of a principal group. Compound **53** is 1H-phenalene-4-carboxylic acid and its isomer **54** is 1H-phenalene-9-carboxylic acid; compound **55** is 2H-pyran-6-carboxylic acid and its isomer **56** is 2H-pyran-4-carboxylic acid; i.e., the pairs of isomers are treated as different acids derived from the same (nonacidic) parent, not as different constitutional isomers of the same acid.

There are other problems for indicated hydrogen, but perhaps enough has been said to show the depth of these waters.

REFERENCES

1. *IUPAC Nomenclature of Organic Chemistry. Definitive Rules for: Section A. Hydrocarbons; Section B. Fundamental Heterocyclic Systems; Section C. Characteristic Groups Containing Carbon, Hydrogen, Oxygen, Nitrogen, Halogen, Sulfur, Selenium and/or Tellurium,* 1969. A, B 3rd Ed.; C, 2nd Ed., Butterworths, London (1971); now available from Pergamon Press, Oxford: (a) pp. 134–139; (b) pp. 186, 187, 190, 193; (c) pp. 262–267; (d) pp. 259–261
2. SCHOENFELD, R., C.S.I.R.O., Melbourne, Australia
3. *IUPAC Nomenclature of Inorganic Chemistry*, 2nd Ed., *Definitive Rules*, 1970, Butterworths, London (1971), p. 21; now available from Pergamon Press, Oxford; also *Pure Appl. Chem.*, **28**, 21 (1971)

7

Stereoisomerism

Stereochemistry is concerned with the arrangement in space of the constituent atoms of molecules, and those features of it that lead to existence of isomers constitute stereoisomerism[1-4]. Many diverse situations are involved but the nomenclature has been greatly simplified by the ability, acquired only recently, to assign absolute configurations[5] and by the promulgation of the sequence rule[6] which permits the names of stereoisomers to be differentiated in most situations in organic chemistry. Methods for handling many of the simpler steric relations have been codified by IUPAC[4] and have proved generally acceptable. However, a new dimension has been introduced by publication[7] of methods recently introduced into *CA* Collective Subject Indexes; its most important features are included in this chapter (*see* p. 140). Full information on this and other aspects of stereochemical nomenclature can be found in the literature cited.

First, however, we should define a few words that are not involved in actual names of compounds but are part of the required vocabulary. There have been variations but we shall follow IUPAC.

The term 'structure' is allotted very wide application by IUPAC – to any aspect of the organization of matter (cf. atomic structure, electronic structure, structure of benzene). 'Constitution' is the term to be used to denote the nature and sequence of bonding of atoms in a molecule.

Compounds having identical molecular formulas but differing in the nature or sequence of bonding or in arrangement of atoms in space are termed 'isomers'. Isomers differing in the nature or sequence of bonding are 'constitutional isomers' (e.g., **1/2**); those differing only in the arrangement of their atoms in space are 'stereoisomers' (e.g., **3a/3b**; **4a/4b**). The adjective 'stereogenic' is applied to an atom or group in a molecule when interchange of two of the atoms or groups attached to it produces a non-identical compound. 'Stereoparent' is the name

$$CH_3CH_2OCH_3$$

1

$$CH_3CH_2CH_2OH$$

2

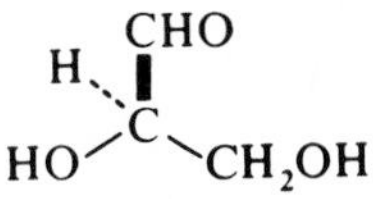

3a

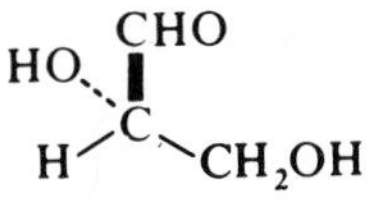

3b

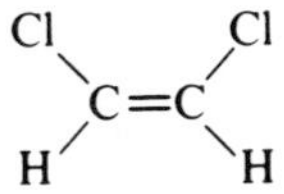

4a

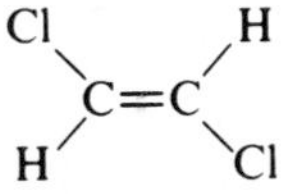

4b

applied to a compound whose wholly or partly trivial name implies the arrangement of some or all of its stereogenic atoms.

Stereoisomers that cannot be converted into their mirror images by only rotation around bonds or translation are optically active; the pairs are termed 'enantiomers'.

This property of non-interconvertibility is now termed 'chirality' (= handedness; from the Greek *cheir* = hand; the 'chi' is pronounced 'ky'); the adjective is 'chiral' and we talk of chiral molecules or chiral groups. Fundamentally, chirality depends on the absence, from a molecule or group, of any element of symmetry except possibly a two-fold axis of rotation (C_2). We may note here, too, for use later, that from a structural point of view a chiral compound may possess one or more centres of chirality and/or axes of chirality and/or planes of chirality[3,6].

Stereoisomers that are not enantiomeric are termed 'diastereo-isomers'; they may be chiral or achiral, as is shown by examples **5–12**, here extracted from IUPAC rule E-4.6. In these formulas, and in general, a thick bond or a wedge (■ or ◀) denotes a bond projecting forwards from the plane of the paper, i.e., towards the observer; a broken bond (- - - or |ıı·) denotes a bond projecting behind the paper, i.e., away from the observer. Normal lines (——) are then considered to be bonds in the plane of the paper. When the spatial arrangement is unknown or unspecified a wavy bond is used (∿)

In Fischer projections, special conventions and restrictions apply. The chiral carbon atom in the centre, in the plane of the paper, is linked at right and left to atoms or groups considered to project toward the observer, and linked above and below to those considered behind the plane of the paper. The principal chain, if there is one, is projected vertically, with the lowest-numbered atom at the top. Thus **11** would appear in Fischer projection as **11A**.

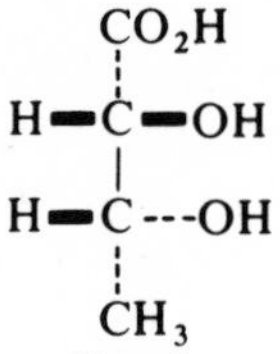

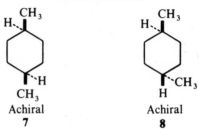

$$
\begin{array}{c}
CO_2H \\
H-C-OH \\
H-C\text{--}OH \\
CH_3
\end{array}
$$

Chiral
5

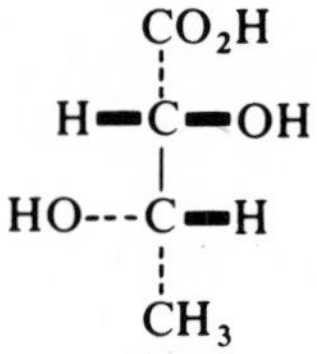

$$
\begin{array}{c}
CO_2H \\
H-C-OH \\
HO\text{--}C-H \\
CH_3
\end{array}
$$

Chiral
6

Diastereoisomers

$$
\begin{array}{c}
CH_3 \\
H\cdots \\
\\
CH_3 \\
H
\end{array}
$$

Achiral
7

$$
\begin{array}{c}
CH_3 \\
H\cdots \\
\\
CH_3 \\
H
\end{array}
$$

Achiral
8

Diastereoisomers

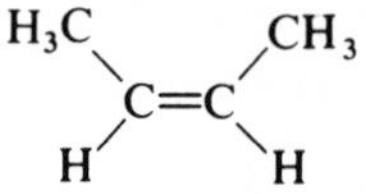

$$
\begin{array}{c}
H_3C \qquad CH_3 \\
C=C \\
H \qquad H
\end{array}
$$

Achiral
9

$$
\begin{array}{c}
H_3C \qquad H \\
C=C \\
H \qquad CH_3
\end{array}
$$

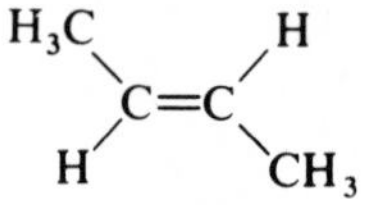

Achiral
10

Diastereoisomers

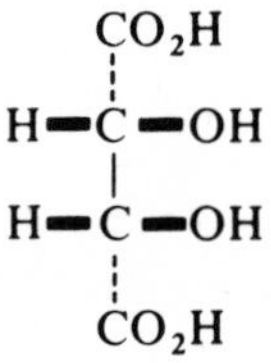

$$
\begin{array}{c}
CO_2H \\
H-C-OH \\
H-C-OH \\
CO_2H
\end{array}
$$

Achiral
11

$$
\begin{array}{c}
CO_2H \\
H-C-OH \\
HO-C-H \\
CO_2H
\end{array}
$$

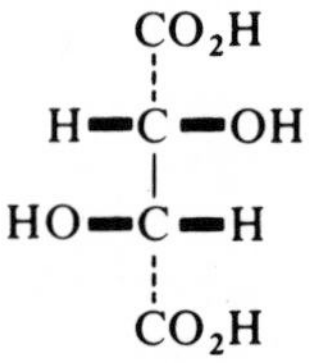

Chiral
12

Diastereoisomers

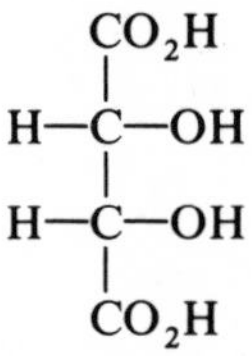

$$
\begin{array}{c}
CO_2H \\
H-C-OH \\
H-C-OH \\
CO_2H
\end{array}
$$

11a

Structure **11** also illustrates a special class of compound — meso compounds — that contain equal numbers of enantiomeric groups identically linked but are themselves achiral.

Stereoisomers that are interconvertible by rotation of atoms or groups around a single bond (or, in the view of some workers[2], also around multiple bonds) are termed 'conformational isomers' (or sometimes 'rotamers' or 'conformers') and those that are conceived as requiring bond-breaking for interconversion are termed 'configurational isomers'; this loose wording is intentional since it has proved impossible so far to draw a sharper boundary[6].

IUPAC Stereochemical Nomenclature

The first IUPAC rule is that stereochemistry is denoted by one or more prefixes to the name of the molecule, or sometimes to constituent groups, and that designation of the stereochemistry must in no other way alter the name or numbering applicable in absence of stereochemical considerations. There are a number of exceptions (stereoparents), where trivial names apply to one stereoisomer, e.g., fumaric acid, or where part or the whole of a complex stereochemistry is specified by a trivial name, e.g., 5β-cholestane or boromycin (18 chiral elements, 262 144 stereoisomers statistically possible)[8]. These principles are accepted also by *CA* except that their inverted indexes are compiled with steric affixes at the end of an entry.

Since the sequence rule procedure is applicable to almost all chiral organic compounds, its elementary principles will be next described.

Sequence Rule Procedure[4,6]

Basic to the sequence rule procedure is the sequence rule itself, by means of which ligands can be arranged in a preferred sequence. Thereafter that sequence can be used in various ways to identify the chiral feature involved.

Let us start with the commonest example, the chiral centre, represented by a central atom with four different ligands a,b,c,d arranged tetrahedrally around it, as shown in **13a** (= **13b**) or **14**. The ligands will be assigned (by the sequence rule) a preferred sequence that can be represented as a > b > c > d, where > denotes 'preferred to'. Clearly this model **13** represents the asymmetric carbon atom so often met in organic chemistry (or any other tetrahedral molecule) and can be converted into its enantiomer **14** by reflexion in a mirror. The sequence rule procedure in this case is to regard the model from

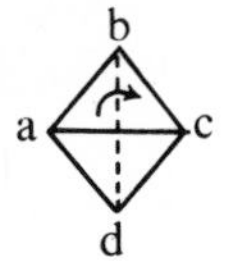

13a   (R)-Form 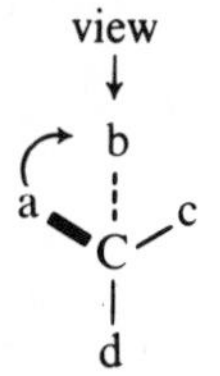   13b 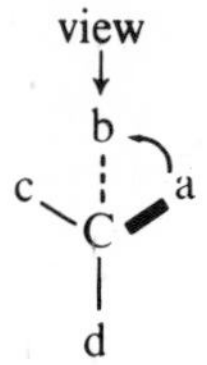   14   (S)-Form

the side remote from the least preferred ligand d, as shown in **13** and
14, and to observe whether the route a–b–c involves a right-handed
(clockwise) or a left-handed (counterclockwise) turn. If right-handed,
as in **13**, that form is termed an R-form (R from the Latin *rectus* =
right); if left-handed, as in **14**, then an S-form (S from the Latin
sinister = left).

The sequence rule symbol R or S, enclosed in parentheses, is placed
in front of the name of the compound, as in **19** and **20**. If several
chiral centres are to be described, the R,S symbols, immediately
preceded by the required locants (no hyphen or space), are collected at
the beginning of the name, as in **30** (*see* p. 137), written as
(1R,3S)-1-bromo-3-chlorocyclohexane.

Next let us imagine that the tetrahedron of formula **13** is elongated
in one direction, as in **15**. This model has an axis of chirality, a twofold
axis, and there is no need for all four ligands to be different to produce

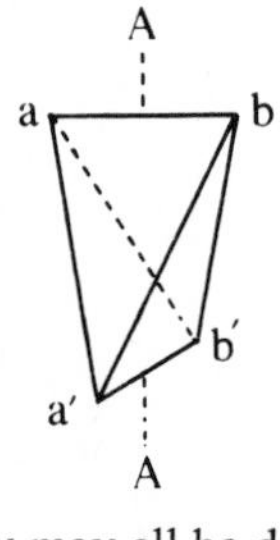

15 A–A = idealized chiral axis

chirality; they may all be different but it suffices that the pair of
ligands a,b at one end of the molecule shall differ from one another,
and that a′,b′ at the other end also differ from one another. In such
cases the sequence rule procedure is to view the model from one end.
It is then considered that the two ligands at that end are preferred to
those at the other. Suppose we view from the top, then we have
a > b > a′ > b′ and thus again a symbol, here R, or aR when necessary
for distinction. As this elongated tetrahedron shares the property of a
screw it is immaterial from which end of the molecule we do our
viewing. Typical compounds having an axis of chirality are **16** and **17**.

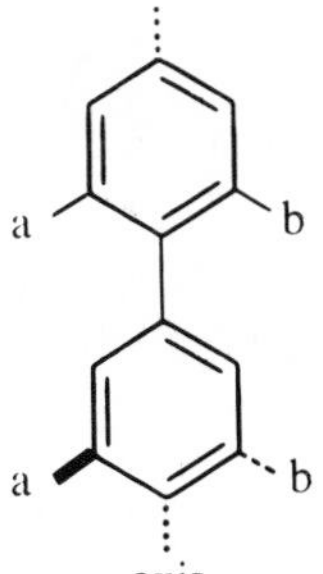

Compounds having a chiral axis

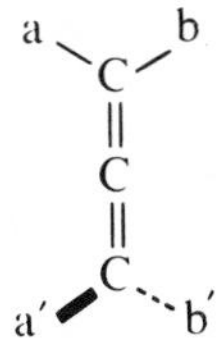

Lastly, compounds having a plane of chirality include those such as
18; here the procedure is to observe the path of the atoms marked
a,b,c from the nearest CH_2 group of the polymethylene chain; the
symbol is R (or pR for distinction) in this case also.

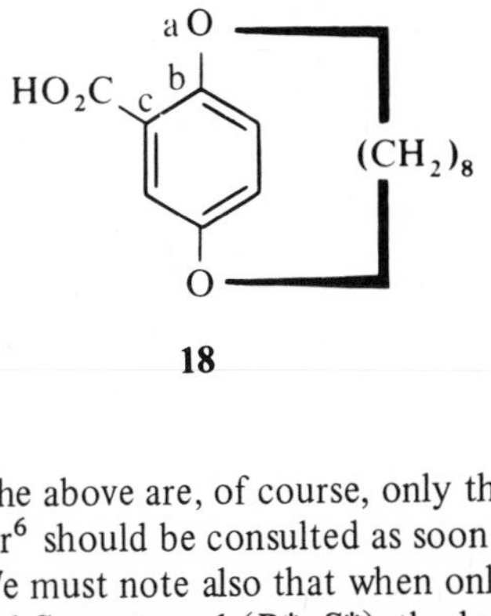

(R)-2,5-(Octamethylenedioxy)-
benzoic acid
An example of a chiral plane

The above are, of course, only the first principles and the original
paper[6] should be consulted as soon as difficulty is perceived.

We must note also that when only relative configurations are known,
R and S are starred (R^*, S^*), the lowest-numbered centre being
assigned R^*. Alternatively the prefix 'rel-' may also be used. A racemate
(phase consisting of equimolar amounts of enantiomers) may be designated (1) by the prefix 'rac-' or (±) plus starred symbols if necessary, or
(2) by (RS) (for a single chiral centre) or (RS) and (SR) (for more than
one chiral centre) plus locants, the lowest-numbered centre always
receiving the (RS) label. When the stereochemistry is wholly or partly
described by a trivial name the prefix 'ent-' may be attached to that
name to designate the enantiomer.

THE ORDERS OF PREFERENCE

The sequence rule itself, which provides an order of preference for ligands, depends on a series of sub-rules, of which the first very widely suffices.

This sub-rule 1 is: Higher atomic number is preferred to lower. It is applied to each of the ligands, working outwards atom by atom along the bonding until all the ligands are arranged in order.

There is clearly no difficulty with **19**; the sequence is Br > Cl > C > H. For **20** we immediately see Cl > C, C > H; the two C's are attached to the central atom and to decide between them we

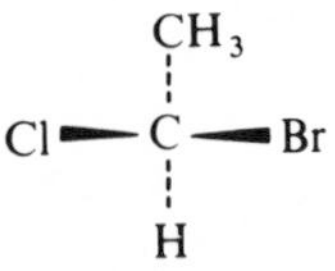

| (R)-1-Bromo-1-chloroethane | (S)-1-Bromo-2-chloropropane |

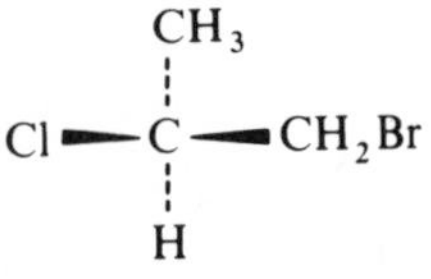

work outwards to the next atoms, which we find to provide Br > H; such situations can be conveniently described as C(Br,H,H) > C(H,H,H). The final result for **20** is thus Cl > CH$_2$Br > CH$_3$ > H.

If we reflect that we have been comparing **A** with **B** we may note that we were comparing branched chains. Branched chains can create much more difficulty but yield to systematic comparisons. For example, in **21** we must proceed from C to N and at each branch decide whether

to take the upper or lower path; the halogen determines that the upper paths be first compared and we find N−C(Cl,H,H) preferred to N−C(H,H,H) (in spite of the bromine!). If we have a case such as **22**, we have CH$_2$Cl as preference at both branches; but as this provides no answer, we proceed to the less preferred branches CH$_2$−CH$_3$ and CH$_3$, with the former clearly preferred.

If an atom is attached to another by a multiple bond, both atoms are regarded as being replicated. The replicates are customarily written in parentheses, as below, to indicate that *only* these atoms are duplicated.

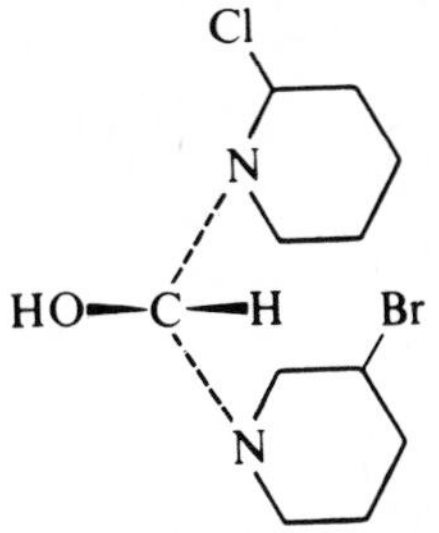

21

(R)-3-Bromo-1-[(2-chloro-1-piperid-
inyl)hydroxymethyl] piperidine

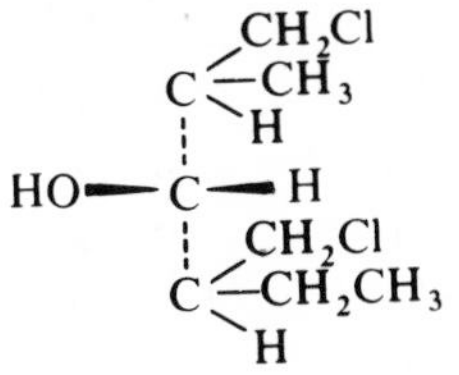

22

(S)-1-Chloro-4-(chloro-
methyl)-2-methyl-
3-hexanol

i.e., that duplication does not apply to the R groups in the alkene
shown if further exploration is required.

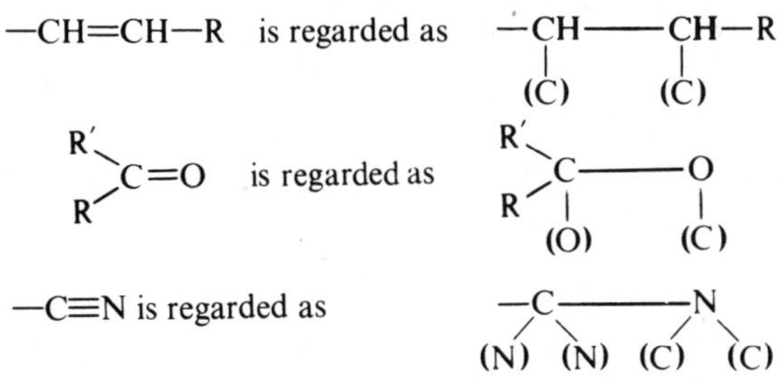

Unsaturated cyclic structures are treated similarly; so both Kekulé
structures afford the same model **23**.

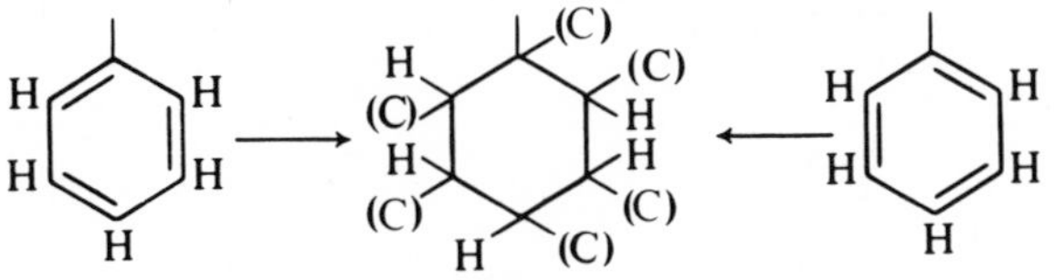

23

Treatment of the Kekulé double bonds
in a phenyl group

Interesting too is that free electron pairs are considered as ligands
of atomic number zero, making, e.g., sulfoxides amenable to the
sequence rule, as in **24**.

$$\text{CH}_3 \diagdown \overset{\displaystyle \cdot}{\underset{\displaystyle \text{C}_2\text{H}_5 \diagup \mathbf{24} \diagdown \text{O}}{\text{S}}}$$

Ethyl methyl (*R*)-sulfoxide (IUPAC)
(*R*)-(Methylsulfinyl)ethane (*CA*)

Many other situations were considered in the original publications, but here only a few more will be mentioned.

The second sub-rule is: Higher mass number is preferred to lower. This enables isotopic labelling to be taken into account, but only if sub-rule 1 has been applied exhaustively and provided no answer.

The third sub-rule deals with situations covered by the classical concepts of *cis–trans*-isomerism which are discussed in the next Section. It is there explained that for isomerism around double bonds this third sub-rule can now be formulated simply as $Z > E$, where the sequence rule is used to decide priorities.

The fourth and fifth sub-rules concern priority between chiral and prochiral centres but their precise formulation is subject to controversy at the time of writing.

cis-trans Isomerism

Different problems arise when a molecule is or can be gainfully regarded as wholly or partly planar. The simplest cases are those involving a double bond or a series of them. There the problem is always to decide which two of the four groups at the ends of the double bond should be selected as determining the well-known *cis* or *trans* prefix. Often this is obvious, as for maleic or fumaric acid, but the number of difficulties was large enough for *CA*[9] to decide in 1968 to use the sequence rule

$$\underset{\textbf{25} \ cis}{\overset{\displaystyle a \diagdown \quad \diagup a'}{\underset{\displaystyle b \diagup \quad \diagdown b'}{\text{C}=\text{C}}}}
\qquad\qquad
\underset{\textbf{26} \ trans}{\overset{\displaystyle a \diagdown \quad \diagup b'}{\underset{\displaystyle b \diagup \quad \diagdown a'}{\text{C}=\text{C}}}}$$

$$\underset{\textbf{27}}{\overset{\displaystyle \text{Cl} \diagdown \quad \diagup \text{Cl}}{\underset{\displaystyle \text{H} \diagup \quad \diagdown \text{CN}}{\text{C}=\text{C}}}}
\qquad\qquad
\underset{\textbf{28}}{\overset{\displaystyle \text{H} \diagdown \ \diagup \text{H}}{\underset{\displaystyle \text{H}_3\text{C} \diagup \ \diagdown \ \text{H} \ \diagup \ \diagdown \text{CO}_2\text{H}}{\text{C}=\text{C}\diagdown\text{C}=\text{C}}}}$$

(*Z*)-2,3-Dichloroacrylonitrile or	(2*E*, 4*Z*)-2,4-Hexadienoic acid
(*Z*)-2,3-Dichloro-2-propenenitrile	

for this purpose. The prefix *Z* (from *zusammen*, German = 'together') was chosen to replace *cis*, as shown in **25**, and *E* (*entgegen*, German = 'opposite') to replace *trans*, as for **26**. Formulae **27** and **28** show specific examples. Example **29** may cause initial surprise.

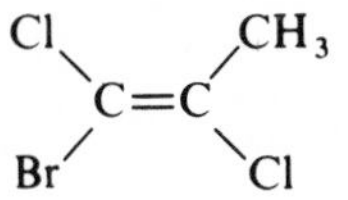

29

(*Z*)-1-Bromo-1,2-dichloropropene

Now *cis*–*trans* prefixes or the sequence rule can obviously be used when two chiral centres exist in a monocyclic system that can be considered as planar (*see* **30**). The *R,S* assignments become, however, extremely cumbrous to assign as the number of chiral centres increases. The IUPAC rules thus prescribe *cis*/*trans* prefixes, abbreviated to *c*/*t* when more than two positions in the ring are substituted, as in **31**; the

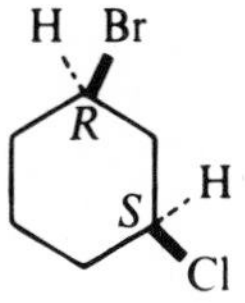

30

(1*R*,3*S*)-1-Bromo-3-chlorocyclohexane

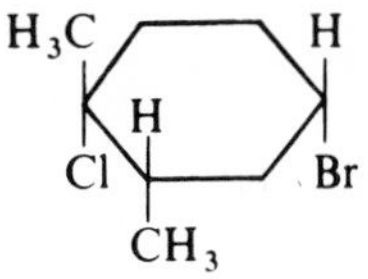

31

c-4-Bromo-1-*r*-chloro-1,*c*-2-dimethylcyclohexane

prefix *r*, applied to the lowest-numbered substituent cited, denotes the reference group to which the *c* and *t* groups are related. *Chemical Abstracts* does not apply these abbreviations and subjects *cis*/*trans* to specific restrictions (*see* pp. 140, 141). *E* and *Z* are not happily applicable to these situations.

All the existing systems become complicated when many similar substituents in the ring are present, yet such compounds are common in nature; wherefore specialist nomenclature has been devised for them, as mentioned on p. 143.

endo, exo, syn, anti

Use of these terms, not included in the IUPAC rules, has been systematized by *CA* and restricted to bicyclo[X.Y.Z] alkanes where

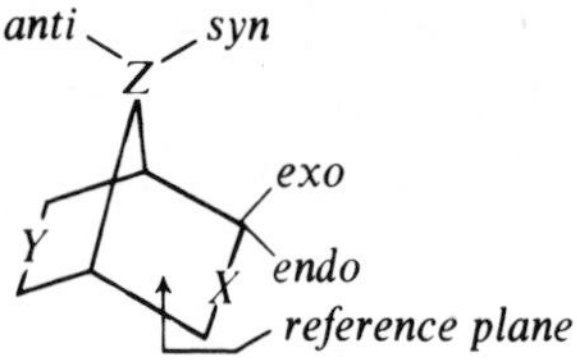

32

$X \geqslant Y > Z$. This use is best explained by the diagram **32**, copied from the *CA* publication[7]; *syn* and *anti* describe the position of substituents on the Z bridge, referred to the X groups; *exo* and *endo* refer to the 'plane' of the molecule including the X and Y groups. These prefixes describe only relative configurations; for absolute configurations see the section on *CA* practices (p. 140).

Stereochemical α,β Infixes

When the structural formula of a bicyclic or polycyclic compound is written in planar form and with a prescribed orientation on the paper, substituents then lying below (behind) the plane of the paper are denoted by broken bonds and called a, and those lying above (in front of) the plane of the paper, denoted by thick bonds, are called β. Such a and β designations are normally attached to and follow the locant of the substituent. The classical case of steroids is discussed briefly on p. 145 but this use of a,β has been partially generalized by IUPAC[10] and has been much extended by the recent *CA* nomenclature described below.

Conformations

Although evidence of separate existence of conformational isomers is variously derived, they cannot as a rule be isolated. The need for prefixes to label individual isomers has therefore been small and only a few are specifically authorized, for describing how groups are attached to six-membered rings[4]. A group so linked by a bond that makes a small angle with the plane containing most of the ring atoms takes the infix (e) (for equatorial) following the group locant, while one linked at a large angle is labelled (a) (for axial), as in **33**. When groups are linked to the carbon atoms of a double bond in a monounsaturated six-membered ring, 'pseudoequatorial' (e') and 'pseudoaxial' (a') are used.

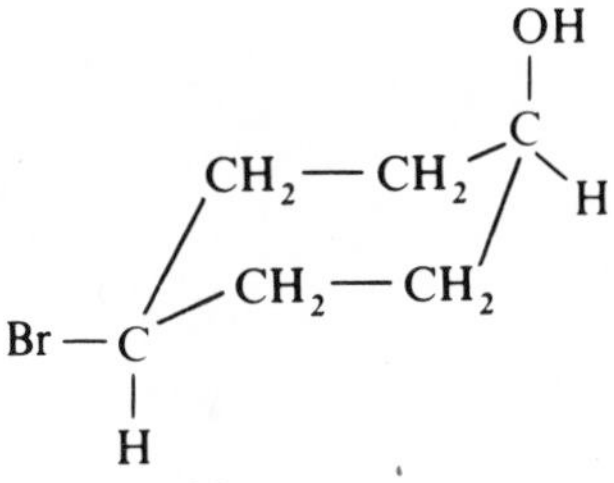

33

4(*e*)-Bromo-1(*a*)-cyclohexanol

Other conformational terms approved by IUPAC[4] but not involved in actual names of compounds include 'chair', 'boat', and 'half-chair' for describing the shape of six-membered rings, and 'eclipsed', 'staggered', etc., to denote the dihedral angle in an A—C—C—B system of bonds. There are also others in more or less common use, particularly in specialist areas such as carbohydrates, polymers, proteins and nucleotides.

Interconversion of conformations depends on rotation around one or more bonds; this can be illustrated for the usual case by rotation around the C—C bond of a substituted ethane as shown in the saw-horse diagrams **34** and **35**. The sequence rule can be employed to name such structures: preferences a,b,c are assigned to each trio of ligands, a and

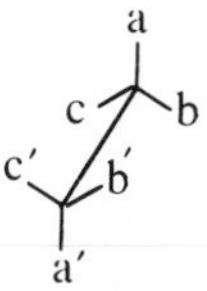

34	**35**	**36**	**37**
Staggered	Eclipsed	*P*-form	*M*-form

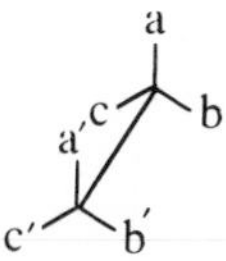

Sawhorse diagrams Newman projections

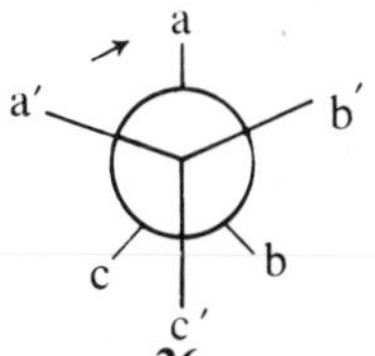

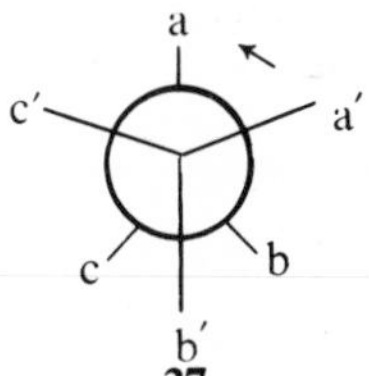

a′ being the sequence-rule-preferred of the trios or the 'odd-man-out' of an abb type. The corresponding Newman projection diagrams **36** and **37** illustrate how rotation of a′ needed to superpose it on a may be right-handed in a *P* form (*P* = plus) or left-handed in an *M* form (*M* = minus). Among the reasons for choice of these prefixes was that + or − can be coupled with numerical statement of the rotation required.

Recent *CA* Index Practice

In 1975 *CA* published[7] revised methods for designation of stereoisomers in its Ninth Collective Index Period (1972–1976); these comprise 16 rules and 30 examples, whose aims are to bring stereoisomers and, in particular, enantiomers, into consecutive entries in their collective indexes and to establish the limits of use of each stereoaffix. Most common situations are covered but only the more important aspects can be illustrated here; reference to the original paper[7] will be rewarding.

In these indexes the stereochemical descriptors are placed after the non-stereochemical names, or after the names of stereoparents; but they are followed by a hyphen which appears to indicate that such descriptors shall precede the rest of the name in cursive English text, as customary hitherto.

The new rules accept previous practice, outlined in preceding sections of the chapter, for the following cases:

(1)   the order of preference of ligands established by the sequence rule;
(2)   the use of *R/S* for statement of chirality when a molecule contains only one tetrahedral centre of known absolute chirality; and
(3)   the *E/Z* method of denoting stereochemistry around a double bond.

However, for most situations a novel procedure is used: relative stereochemistry is cited by one or more of the affixes R^*/S^*, a/β, *cis/trans*, or *endo/exo*, applied as described earlier in this Chapter and enclosed in parentheses ( ). The absolute chirality, if known, is then recorded as *R* or *S* for the lowest-numbered or, sometimes, the sequence-rule-preferred chiral centre, that symbol being preceded by its locant (when required); this absolute descriptor is placed before the relative

$$H_3C \text{---} \overset{\displaystyle Cl}{\underset{\displaystyle H}{C}} \text{---} \overset{\displaystyle H}{\underset{\displaystyle OH}{C}} \text{---} CH_3$$

38

[*S*-(*R**,*R**)]-3-Chloro-2-butanol

descriptor and separated from it by a hyphen, the whole being then placed in square brackets, e.g., [1*R*-(1*a*,3*a*,5*β*)] -. The following examples are taken or adapted from the *CA* publication[7].

When there are only two chiral centres in a chain, as in **38**, with only the relative configuration known, this is expressed as R^*,R^* for centres of identical chirality or $R*,S*$ for those of unlike chirality. Then, when the absolute chirality is known, this is added by R or S determined for the chiral centre preferred by the sequence rule when necessary.

Somewhat similar principles apply when more than two chiral centres are present in a chain, but then locants must be attached to the symbols in brackets, as in the name of **39**. For relative chirality the lowest-numbered chiral centre receives the label $R*$ and the others are designated by their steric relation to that centre. If absolute chirality is known, then the lowest-numbered centre receives R or S according to its actual chirality, placed as shown in **39**.

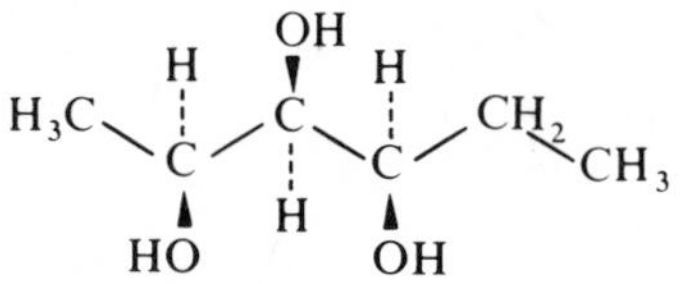

39
[2S-(2R*,3S*,4S*)]-2,3,4-Hexanetriol

In the complete name for **39** it appears at first sight that S and $R*$ should not both refer to the same centre 2. However, in the name for the enantiomer of **39**, the descriptors are [2R-(2R*,3S*,4S*)]-, and the apparent anomaly vanishes.

The *CA* use of *exo*, *endo*, *syn*, and *anti* is sufficiently explained by **40**; in this the locants for *endo* are not required. Note, however, that the system is inapplicable when such compounds contain additional

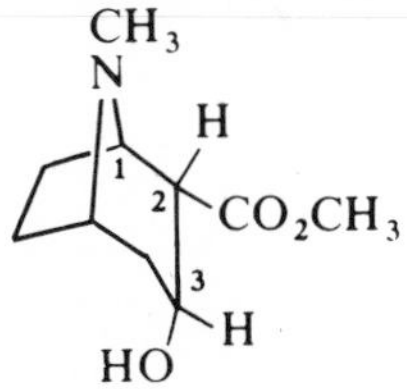

40 Methyl [1R-(endo,endo)]-3-hydroxy-8-methyl-8-aza-
bicyclo[3.2.1]octane-2-carboxylate

rings attached to the bicyclo[X.Y.Z] alkane skeleton; for such systems *CA* uses α and β as described below or reverts to an entry 'stereoisomer'.

When only two chiral centres are present in an eight-membered or smaller ring or in a ring system consisting of such rings, *cis* or *trans* is used as the relative descriptor; *see* **41**.

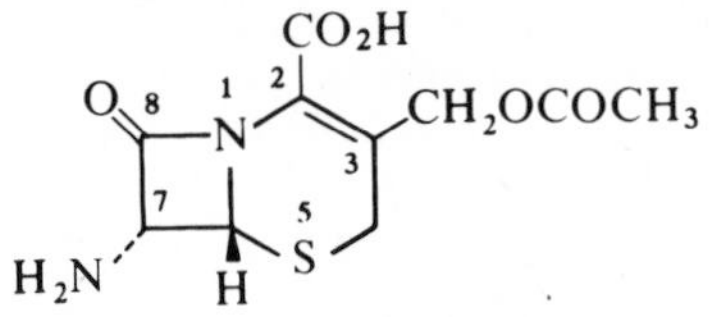

41   (6*R-trans*)-3-[(Acetyloxy)methyl]-7-amino-8-oxo-5-thia-1-
azabicyclo[4.2.0]oct-2-ene-2-carboxylic acid

However, when more than two chiral centres are present in an eight-membered or smaller ring or in a ring system containing such rings, a and β are used to describe the relative stereochemistry; R or S, plus a locant, for the lowest-numbered chiral centre provides the absolute stereochemical descriptor as usual.

Let us work out step by step a common type of problem, namely the systematic nomenclature for a compound whose formula **42** is presented as its absolute configuration. First we must denote the relevant stereochemistry at each stereogenic centre, for inclusion in parentheses; we note that 8a-CH_3 is on the same side as the 'plane' of the molecule as CO_2H, whilst 4a-H and OH are on the opposite side; the lowest-numbered stereogenic centre is C-2, so that the 2-CO_2H receives the designation a, and then the relative designation for the whole molecule is (2a,4aβ,6β,8aa). Finally to convert this relative designation into the absolute designation we note that the absolute stereochemistry at the lowest-numbered stereogenic centre is S, so that the total absolute designation is [2S-(2a,4aβ,6β,8aa)]-. The situation noted for compound **39** is met here again, namely that the expression in parentheses is exactly the reverse of that shown in the formula; but again this is coincidental, for there would have been agreement if we had worked on the enantiomer of **42**.

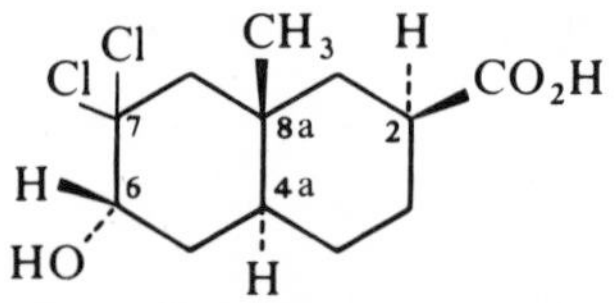

42   [2S-(2a,4aβ,6β,8aa)]-7,7-Dichlorodecahydro-6-hydroxy-8a-
methyl-2-naphthalenecarboxylic acid

a and β are also often used to identify the chirality of additional centres not covered by the name of a cyclic stereoparent. These parents usually have known absolute stereochemistry and in example **43** (taken from the *CA* paper) the 5a,17a in parentheses also refer to absolute stereochemistry. This cannot be altered since it has been part of the

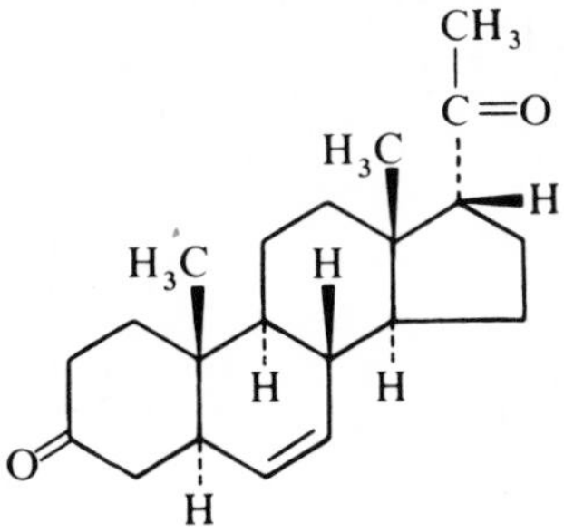

43   (5α,17α)-Pregn-6-ene-3,20-dione

enormous steroid literature over several decades, and it is therefore of great importance to realize that the generalized use of a/β, as in **42**, denotes only relative stereochemistry. It is further of equal importance to remember that use of a and β for stereoparents requires orientation of the formula on paper with one specific face towards the observer; if that is not done, then a and β are interchanged. In *CA* indexes this will be established by their conventional rules; it may not be so easy for the experimental chemist.

  R^* and S^* are also often used as relative descriptors in a number of less common situations, but the original paper should be consulted for these and other points[7].

Specialist Areas

While it is a virtue of the sequence rule procedure that it takes no account of the formal classification of compounds or of their biogenetic relations or mechanistic interpretations, yet its results can be most inconvenient when, in a related series of compounds, formulas that are clearly analogous receive different stereochemical designations. There are, therefore, several fields — very important ones — for which more convenient nomenclatures have been developed and are now always used.

CARBOHYDRATES

The oldest of these other systems is the use of D and L (originally italic capitals, now small capitals) for sugars according as the substituent on the highest-numbered asymmetric carbon atom is on the right or on the left of the classical Fischer projection illustrated in the generalized open-chain form for a D-sugar **44**. Originally formulated as a pure convention for D-glyceraldehyde, this proved correct for absolute configurations[5] and is now applied to the whole field of carbohydrates and their derivatives[11].

The full nomenclature of a monosaccharide is exemplified for β-D-glucopyranose **45** as follows: pyranose denotes the six-membered ring (pyran) of a sugar ('-ose'); gluco specifies the relative configurations

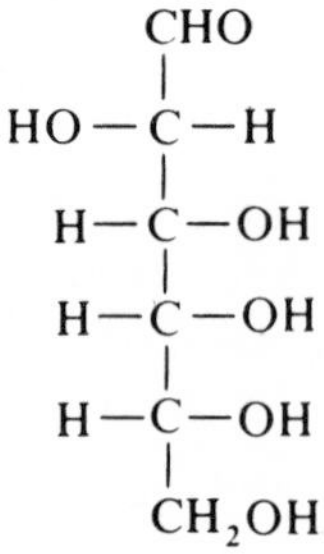

CHO
|
HO — C — H
|
H — C — OH
|
H — C — OH
|
H — C — OH
|
CH$_2$OH

44

Linear form of
a D-hexose

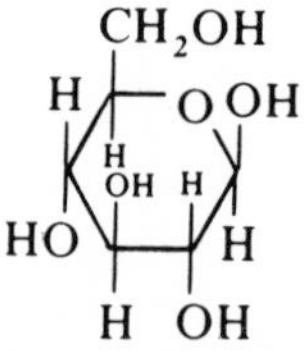

CH$_2$OH

H ┬ O OH
  H
  OH H
HO └──┴ H
H OH

45

Haworth projection of
β-D-glucopyranose

at C-2, C-3, C-4, and C-5; β denotes the stereochemistry at C-1; and D identifies this rather than the enantiomeric form of the whole molecule.

R/S symbolism is used for chiral elements not included in the main carbohydrate chain, e.g., *O,O′*-alkylidene groups RHC< attached to sterically non-equivalent oxygen atoms.

AMINO ACIDS

Early work on amino acids and proteins was approximately contemporaneous with that on sugars; the configuration was based on **46**, R = CH$_2$OH, for L-serine. Since, however, natural amino acids are α-amino compounds, the D/L here referred to the lowest-numbered

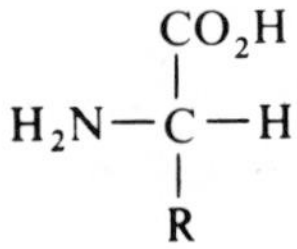

CO$_2$H
|
H$_2$N — C — H
|
R

46   Any L-α-amino acid

asymmetric atom, and confusion with carbohydrate nomenclature (based on the highest-numbered asymmetric atom) was inevitable. Temporary conventions could, however, be abandoned when again the convention for L-serine proved correct in an absolute sense, so that use of R/S became possible here too and is now general. Specialists have penetrated deeply into the stereochemistry of polypeptides and proteins; the resulting complex nomenclature can be found in the literature[12].

STEROIDS

Steroids occupy a further large field where specialist nomenclature was essential[13]; the resulting nomenclature principles have proved valuable also in other sectors.

The tetracyclic cyclopentaphenanthrene skeleton common to almost all steroids can be regarded as planar and oriented as shown in **47**; groups or atoms attached to this system can then be labelled a or β

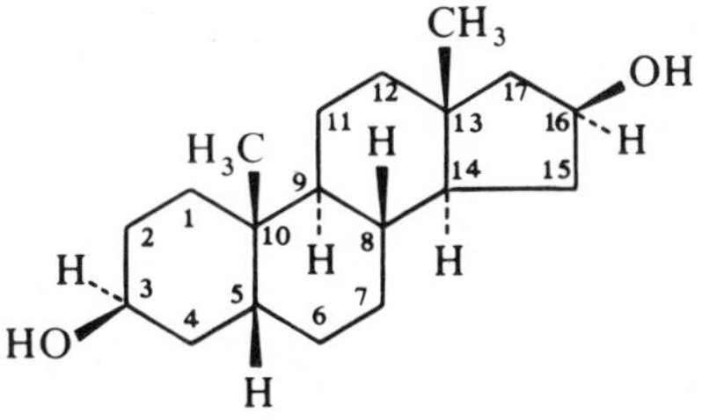

47 5β-Androstane-3β,16β-diol

according to whether they are, respectively, behind or in front of the plane of the paper. These a/β assignments are here absolute because, for the third time in an important field, the conventional stereochemistry of the pioneers proved fortuitously to be 'absolutely' correct. A simple case will show the principles: In the name 5β-androstane-3β,16β-diol for compound **47**, the syllables 'androstane' denote a parent compound comprising a cyclopentaphenanthrene nucleus, fully hydrogenated, with two methyl groups (at positions 10 and 13) as sole substituents, and with the stereochemistry shown at positions 8, 9, 10, 13, and 14; 5β is specified separately because there are many derivatives with H-5 missing or replaced by 5a; the '-3β,16β-diol' states the additional groups.

In the voluminous chemistry of steroids are many varieties of structure, involving chains attached to the tetracyclic nucleus, others having additional ring systems, the latter in many cases spiro-attached so that the new ring is no longer in the same plane as the tetracyclic system. For chirality in these chains and spiro-attached components recourse is had to the *R/S* system. Very many such compounds have been allotted stereoparent names, the fixed symbolism applying to both *a/β* and *R/S*. Hecogenin (48) is one such compound; the formula 49 represents (22*S*)-5a-solanidane, one of the considerable number of alkaloids having a steroid basis in structure and symbolism.

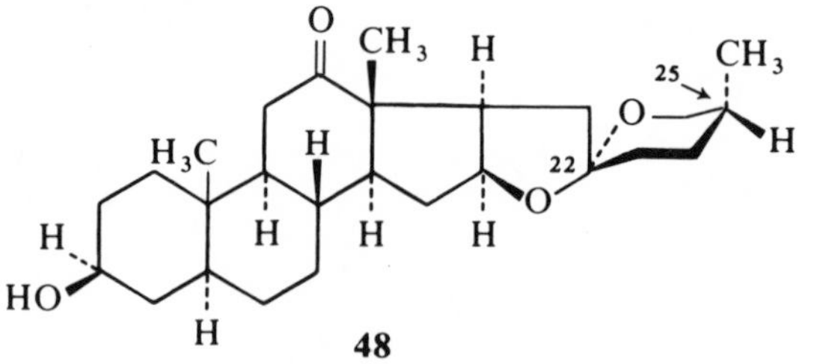

48

Hecogenin

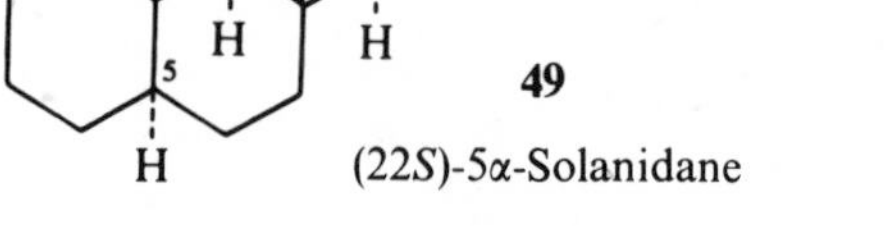

49

(22*S*)-5α-Solanidane

(According to the *CA Index Guide* all the stereogenic centres except C-5 and C-22 are comprised within the stereoparent name)

As in all series, stereochemistry implicit in the parent names can be modified by stating the divergent feature in a prefix. For example, 5a,10a-androstane has the 10-methyl group *a*-oriented; the prefix '10ξ-' in 5a,10ξ-androstane shows that the orientation at position 10 is in this case uncertain; 25*S*-hecogenin has the alternative orientation at position 25.

Finally, the full panoply of auxiliary symbolism finds use among steroids: *R**/*S**, *rel*, *ent*, *rac*, (±) (*see* p. 133).

CYCLITOLS

Cyclitols, i.e., cycloalkanes containing one hydroxyl group on each of
three or more ring atoms, appear related to carbohydrates by virtue of
the multiplicity of hydroxyl groups but in fact this feature brings a
need for special numbering and special treatment of stereochemical
nomenclature. Several early methods and the R/S system proved
unsuitable for general use, so a novel system was adopted by IUPAC
and seems satisfactory[14].

Projection formulas for cyclitols are normally drawn as in **50**, i.e.,
with the cyclohexane ring on its side and hydroxyl groups represented

50

by vertical lines. This drawing makes it easy to divide the hydroxyl
groups into two sets, one set above the plane of the ring (three hydroxyl
groups in **50**) and the other (one such group in **50**) below. Lowest numbers are assigned to one of these sets according to rules devised by
IUPAC/IUB, whose publication should be consulted for details.
Numbering conventionally begins across the top bond of the ring; this
may require redrawing the structure. Then relative configurations are
described by writing the series of numbers comprising the two sets like
a fraction, e.g., 1,2,4/3 for **50**. However, *CA* prefers to describe such
configurations (except for inositols; *see* below) by the α/β system used
for steroids (*see* p. 145).

To specify absolute chirality, formulas can be written vertically as
51A and **51B**; then if the 1-OH bond points to the right, as in **51A**,

51A	**51B**
1L-1,2,4/3-Cyclohexanetetrol	1L-1,2,4/3-Cyclohexanetetrol

the D prefix is used; if to the left, as in **51B**, the L prefix. Alternatively,
it will be found that in formulas drawn with the ring on its side, the
sequence of numbering is counterclockwise in D- but clockwise in
L-compounds.

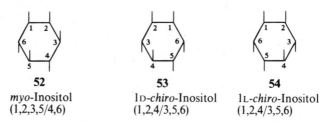

52	53	54
myo-Inositol	1D-*chiro*-Inositol	1L-*chiro*-Inositol
(1,2,3,5/4,6)	(1,2,4/3,5,6)	(1,2,4/3,5,6)

The nine cyclohexanehexols are given the special name 'inositols' by both IUPAC/IUB and *CA*. These compounds are differentiated by italicized, hyphenated prefixes, as in *myo*-inositol, **52**. Of these nine isomers, only two are chiral and they are enantiomers, **53** and **54**, known earlier simply as D- and L-inositol, but now as 1D- and 1L-*chiro*-inositol.

LIPIDS

Lipids comprise esters or ethers of glycerol, often as relatively small components of larger molecules. They are of great concern to biochemists as well as to organic chemists, and in retrospect it is no surprise that, as excellently discussed in the relevant IUPAC/IUB report[15], the earlier special nomenclature devised for this stereochemistry proved inadequate for the demands of biochemists as well as for the variety of closely related derivatives. Four nomenclature systems, including the *R/S* system, had to be abandoned, and a new system, pioneered by Hirschmann[16], was adopted by IUPAC/IUB; it is known as the '*sn-*' system (*s*tereospecific *n*umbering) and proved adequate to deal with migration or interchange of *O*-substituents and with the potentialities of enzymes.*

According to the *sn*-system, glycerol should be arranged as a Fischer projection written as **55**, i.e., with the central OH group to the left *in all cases*, including all derivatives. The glycerol carbon atoms are then numbered vertically downward as 1, 2, 3. The effect of this can be sufficiently demonstrated by considering the familiar L-*α*-glycerophosphoric acid which, being written as **56**, becomes numbered as shown and named *sn*-glycerol 3-(dihydrogen phosphate) or *sn*-glycero-3-phosphoric acid. Then note that **57** is the formula of *sn*-glycerol 1-(dihydrogen phosphate) or *sn*-glycero-1-phosphoric acid, and that this is the *enantiomer* of **56** (**56** and **57** are not superposable

---

*Owing to prochirality the two CH_2OH groups of glycerol are enzymically distinguishable. This is an area where the sequence-rule procedure is in dispute.

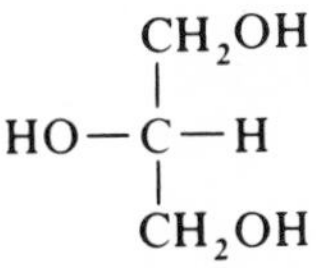

$$\underset{\textbf{55}}{\begin{array}{c} CH_2OH \\ | \\ HO-C-H \\ | \\ CH_2OH \end{array}} \qquad \underset{\textbf{56}}{\begin{array}{c} ^1CH_2OH \\ | \\ HO-^2C-H \\ | \\ ^3CH_2OPO_3H_2 \end{array}} \qquad \underset{\textbf{57}}{\begin{array}{c} CH_2OPO_3H_2 \\ | \\ HO-C-H \\ | \\ CH_2OH \end{array}}$$

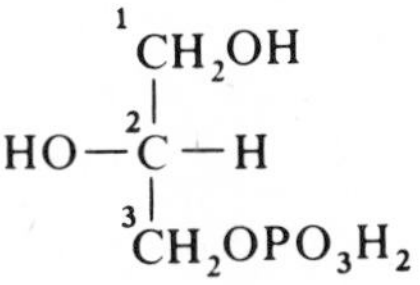

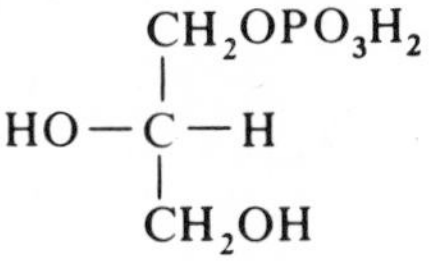

and are interconverted by interchange of any two ligands). No other indication of this relationship is given, which appears contrary to the main principle adopted for *CA* indexes.

ORGANOMETALLIC COMPOUNDS

For stereoisomerism in this area see p. 35 and the references quoted there.

A FINAL WORD

A completely different approach to stereochemistry by Prelog[8] is based on combinations of simplexes, considered two-dimensionally and three-dimensionally; although of fundamental importance for interpretation of stereoisomerism its arguments have not yet been incorporated into a systematic nomenclature.

REFERENCES

1. For a short introduction *see* ELIEL, E.L. and BASOLO, F., *Elements of Stereochemistry*, John Wiley and Sons, New York-London-Sydney-Tokyo (1969)
2. For a stimulating short text *see* MISLOW, K., *Introduction to Stereochemistry*, W.A. Benjamin, 2nd Ed., Reading, Mass. (1978)
3. For a crystallographic and point-group discussion *see* JAFFÉ, H.H. and ORCHIN, M., *Symmetry in Chemistry*, John Wiley and Sons, Inc., New York-London-Sydney (1965)
4. IUPAC COMMISSION ON NOMENCLATURE OF ORGANIC CHEMISTRY, *Rules for the Nomenclature of Organic Chemistry, Section E: Stereochemistry*, 1974, Pergamon Press, Oxford-New York-Paris-Frankfurt; *Pure Appl. Chem.*, **45**, 11 (1976)
5. BIJVOET, J.M., PEERDEMAN, A.F. and VAN BOMMEL, A.J., *Nature*, **168**, 271 (1951); *cf.* BIJVOET, J.M., *Proc. Kon. Ned. Akad. Wet. Amsterdam*, **52**, 313 (1949)
6. CAHN, R.S., INGOLD, SIR C. and PRELOG, V., *Angew. Chem. Internat. Ed.*, **5**, 385 (1966); for abbreviated accounts *see* ref. 4, appendix 1, or CAHN, R.S., *J. Chem. Educ.*, **41**, 116, 503 (1964)
7. BLACKWOOD, J.E. and GILES, P.M., Jr., *Chemical Abstracts Stereochemical*

Nomenclature of Organic Substances in the Ninth Collective Period (1972–1976), *J. Chem. Inform. Computer Sci.*, **15**, 67 (1975); *Index Guide to Vol.* 76 (1972) or *Ninth Collective Index, Vols.* 76–85 (1972–1976), *Index Guide*, para. 203

8. PRELOG, V., Nobel Lecture 'Chirality in Chemistry', The Nobel Foundation, Stockholm, 1976. Reprinted in *J. Mol. Catalysis*, **1**, 159–172 (1975/76), and elsewhere

9. BLACKWOOD, J.E., GLADYS, C.L., LOENING, K.L., PETRARCA, A.E. and RUSH, J.E., *J. Am. Chem. Soc.*, **90**, 509 (1968); BLACKWOOD, J.E., GLADYS, C.L., PETRARCA, A.E., POWELL, W.H. and RUSH, J.E., *J. Chem. Docum.*, **8**, 30 (1968)

10. IUPAC COMMISSION ON NOMENCLATURE OF ORGANIC CHEMISTRY, *Nomenclature of Organic Chemistry: Section F: Natural Products and Related Compounds (Provisional)*, IUPAC Information Bull., Appendix No. 53 (1976)

11. IUPAC/IUB, *Carbohydrate Nomenclature-1 (Tentative Rules)* (1969), *Biochem. J.*, **125**, 673 (1971); *Biochemistry* **10**, 3983 (1971)

12. IUPAC/IUB, *Nomenclature of α-Amino Acids; Biochemistry*, **14**, 445 (1975); *Eur. J. Biochem.*, **53**, 1 (1975)

13. IUPAC/IUB, *Nomenclature of Steroids (Rules* 1971); *Pure Appl. Chem.*, **31**, 283 (1972)

14. IUPAC/IUB, *Nomenclature of Cyclitols (Recommendations* 1973), *Pure Appl. Chem.*, **37**, 283 (1974)

15. IUPAC/IUB, *The Nomenclature of Lipids (Recommendations* 1976); *Hoppe-Seylers Z. Physiol. Chem.*, **358**, 617 (1977); *Mol. Cell. Biochem.*, **17**, 157 (1977)

16. HIRSCHMANN, H., *J. Biol. Chem.*, **235**, 2762 (1960)

8

Natural Products

The field of natural products is an old and enormous one. We here
restrict the term to organic chemistry but then use it in the broad sense
for compounds extracted from bodies of plants, animals, or micro-
organisms.

Recent IUPAC general recommendations[1] mostly record practices
already used by careful chemists. A new natural product is usually
assigned a trivial name that serves to represent it until its structure is
known, and is often used thereafter if the systematic name proves long
and cumbersome. By usage the trivial name is derived from that of the
source material: e.g., nicotine from *Nicotiana tabacum*, and formic acid
from ants (*Formica*). In these examples the function of the compound
is correctly expressed by the endings; when such function is not yet
known, the name should *not* have an ending having structural implica-
tions. So many endings are already in such use that IUPAC felt it
desirable to propose the noncommital but strange and unaesthetic
suffixes '-un', '-une', and '-ur' (urgh!) for new trivial names. Whether
these will be accepted instead of the long-used '-in' (insulin, penicillin,
inulin, etc.) remains to be seen.

A compromise between structurally uninformative biologically based
short names and systematic long complex ones is provided by semisystematic names based on defined parent structures. Those recommended by IUPAC, which *CA* calls Class B stereoparents, are usually derived
for families of natural products by removing all functional groups,
retaining hydrocarbon groups that produce chiral centres at the
skeleton, and considering the skeleton as fully hydrogenated. Thus
the prostaglandins, exemplified by **1**, are considered to be derived from
prostane, **2**, wherein the stereochemistry shown by the structure and
the systematic name is implied. Substituents and other modifications
in the parent structure are then designated by prefixes and suffixes,
and their positions by locants, in the usual manner. It should be

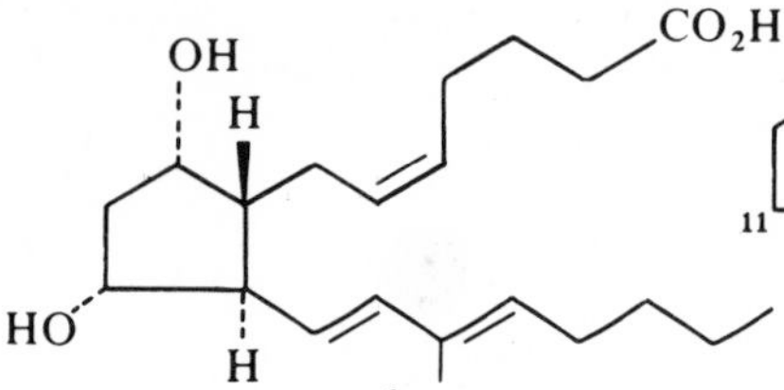

1

(5Z,9α,11α,13E,15E)-9,11-
Dihydroxy-15-methylprosta-
5,13,15-trien-1-oic acid

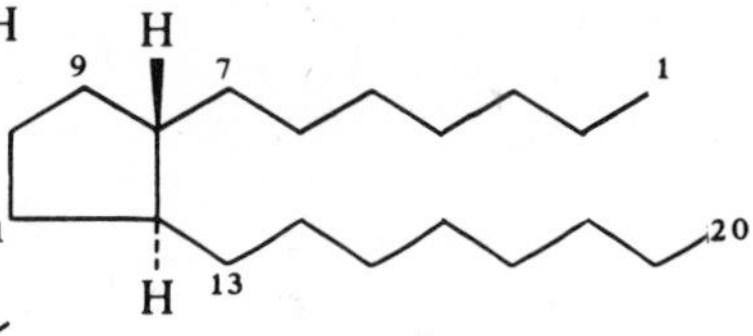

2

Prostane
(1S-*trans*)-1-Heptyl-2-
octylcyclopentane

emphasized that the stereoparent name implies the absolute configuration at all chiral centres present.

For natural products of known structure that are not members of a substantial family, *CA* treats the trivial name of a common one as a Class C stereoparent. Such a name implies both absolute stereochemistry and function. It may be modified as usual, even with prefixes for functional groups that normally have priority, i.e., are cited as suffix. Some random examples of Class C stereoparents from *CA* indexes (which may be consulted if structures are of interest) are rifamycin, corynan, and vincaleukoblastine.

Many of the IUPAC/IUB tentative rules and recommendations in following sections of this chapter are available in a compilation with coverage up to 1978[2].

Carbohydrates

PARENT NAMES AND STEREOCHEMISTRY

Fischer projection formulas in which the carbonyl carbon atom is at or near the top of the structure are easiest to use for associating names and formulas for simple sugars (monosaccharides), although Haworth representations are also permissible. Trivial (stereoparent) names are preferred for the C_3 to C_6 (IUPAC/IUB[3]) or C_5 to C_6 (*CA*[4a]) monosaccharides (e.g., arabinose, ribose, galactose, glucose, mannose, fructose), but systematic names may be formed for these and must be used for larger ones. A systematic name is formed from

(1)   one or more prefixes describing configuration,
(2)   the numerical syllable(s) citing the number of carbon atoms in the chain, and
(3)   the ending '-ose' (for aldoses) or '-ulose' (for ketoses).

Each compound prefix consists of an italicized aldose name minus the '-se' ending (*erythro-*, *ribo-*, *gluco-*, *manno-*, etc.), which describes the relative configuration of four or fewer chiral carbon atoms of the chain, and a preceding D or L that cites the absolute configurations of the highest-numbered chiral carbon atom, as noted on p. 143.

Thus are obtained names such as D-*arabino*-2-hexulose (for D-fructose) and D-*glycero*-D-*gluco*-heptose (note that two configurational prefixes are sometimes needed). Configurations are further

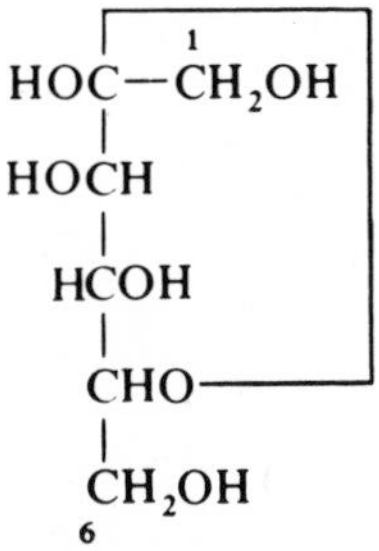

$$
\begin{array}{l}
\text{HOC}-\text{CH}_2\text{OH} \quad ^1 \\
\text{HOCH} \\
\text{HCOH} \\
\text{CHO} \\
\text{CH}_2\text{OH} \quad _6
\end{array}
$$

3
β-D-Fructofuranose
β-D-*arabino*-2-Hexulofuranose

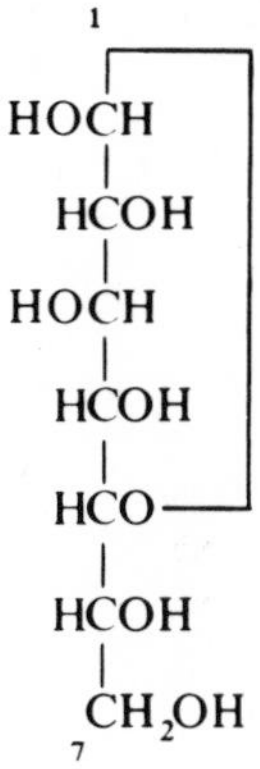

$$
\begin{array}{l}
^1 \\
\text{HOCH} \\
\text{HCOH} \\
\text{HOCH} \\
\text{HCOH} \\
\text{HCO} \\
\text{HCOH} \\
\text{CH}_2\text{OH} \quad _7
\end{array}
$$

4
β-D-*glycero*-D-*gluco*-
Heptopyranose

specified for the highest-numbered chiral carbon atom by the prefixes D and L and for the glycosidic carbon atom (i.e., the one with two bonds to oxygen) by the prefixes a and β, as in the common a-D-glucose. Special terminations are used for dialdoses ('-odialdose'), aldoketoses ('-osulose'), and diketoses ('-odiulose'). The acyclic forms of sugars are designated by the prefix 'aldehydo-' or 'keto-', and the ring size in cyclic ones with the endings '-furanose' or '-pyranose'. Two of the resulting names are shown in **3** and **4**.

SUBSTITUTED FORMS

Substitution of hydrogen on oxygen, or of the carbonyl oxygen, is mostly cited as usual in organic nomenclature, e.g., 3-*O*-methyl-D-glucose or D-glucose 3-methyl ether; 2,4-di-*O*-acetyl-*aldehydo*-D-mannose or *aldehydo*-D-mannose 2,4-diacetate **5**; 1,2-*O*-isopropylidene-a-D-glucofuranose **6** [IUPAC/IUB; *CA* calls the bivalent radical

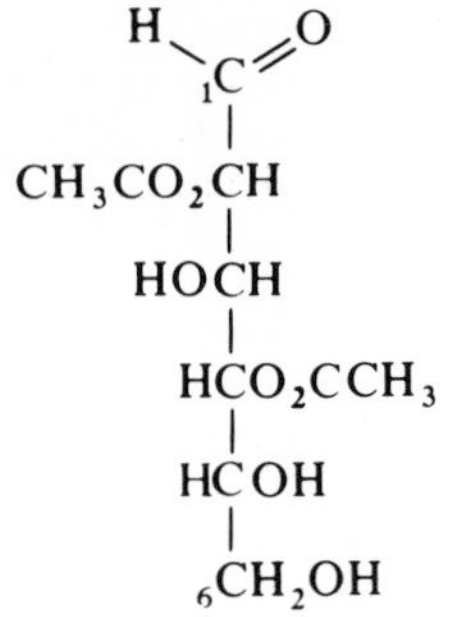

$$\begin{array}{c} \overset{H}{\underset{1}{\diagdown}}\underset{|}{C}{\diagup}^{O} \\ CH_3CO_2\underset{|}{CH} \\ HO\underset{|}{CH} \\ H\underset{|}{C}O_2CCH_3 \\ H\underset{|}{C}OH \\ _6CH_2OH \end{array}$$

5

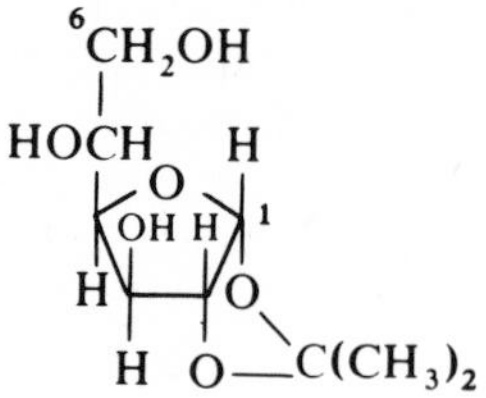

6

(1-methylethylidene)] . However, alkyl or aryl substitution on the glycosidic hydroxyl group leads to glycosides, and 'alkyl (or aryl) -oside' names, such as methyl β-D-glucoside, are used for these mixed acetals. Similarly 1-*C*- sugar radicals are called glycosyl radicals, as in a-D-mannosyl, and the related 1-*O*- ones glycosyloxy (illustrating the R- alkyl, RO— alkoxy relationship). Replacement of a hydroxyl group (other than at C-1) by hydrogen is described by a locant and the prefix

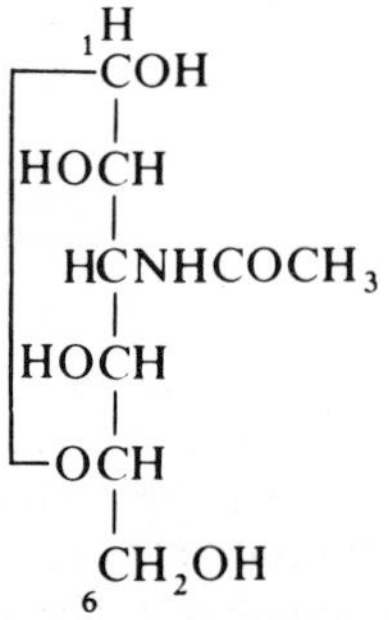

$$\begin{array}{c} \overset{H}{\underset{1}{-}}\underset{|}{C}OH \\ HO\underset{|}{CH} \\ H\underset{|}{C}NHCOCH_3 \\ HO\underset{|}{CH} \\ O\underset{|}{CH} \\ \underset{6}{CH_2OH} \end{array}$$

7

3-Acetamido-3-deoxy-α-L-glucopyranose
3-(Acetylamino)-3-deoxy-α-L-glucopyranose (*CA*)

'deoxy-'; if such hydrogen is then itself replaced by another group, the usual substitutive prefix is used and the locant repeated, as in 7. An 'anhydro-' prefix is used to denote elimination of water between two specified hydroxyl groups: 2,3-anhydro-4-*O*-methyl-a-D-mannopyranose.

REDUCED AND OXIDIZED PRODUCTS

Polyhydric alcohols corresponding to aldoses are named by changing the '-ose' of the aldose name to '-itol', as in D-mannitol. If a saccharide

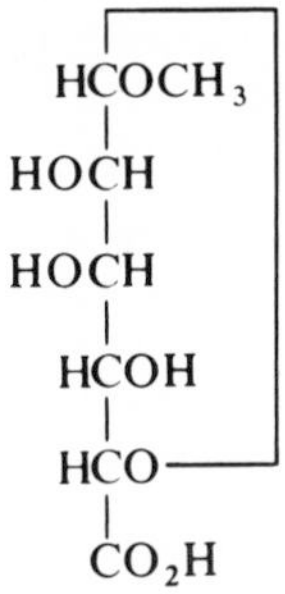

$$
\begin{array}{c}
\text{HCOCH}_3 \\
|\\
\text{HOCH} \\
|\\
\text{HOCH} \\
|\\
\text{HCOH} \\
|\\
\text{HCO} \\
|\\
\text{CO}_2\text{H}
\end{array}
$$

8

Methyl α-D-mannopyranosiduronic acid

be represented by $HOCH_2$ - - - CHO, the ending for the $HOCH_2$ - - - CO_2H derivative is '-onic acid', that for HO_2C - - - CHO '-uronic acid', and that for HO_2C - - - CO_2H '-aric acid': D-gluconic acid, L-glucaric acid, methyl α-D-mannopyranosiduronic acid **8**.

OLIGOSACCHARIDES (SMALL POLYMERS)

Common disaccharides are still called by their old trivial names: sucrose, maltose, lactose, etc. However, they may be named systematically (and are so named by *CA*) as glycosyl glycosides (non-reducing) or glycosylglycoses (reducing), e.g., β-D-fructofuranosyl α-D-glucopyrano-side (sucrose). In trisaccharides the positions involved in linking the two glycosyl groups are cited by the two locants and one or two arrows showing the sequence, as suggested by - - - osyl-(1→4) - - - osyl-(1→4) - - - ose.

Cyclitols

Cyclitol nomenclature[4b,5] presents few special features beyond those already discussed under stereochemistry (*see* p. 147). Inositol names are used as stereoparents by *CA*, which also names cyclohexanepentols as deoxyinositols, e.g., **9**.

Substituents on oxygen (and for inositols, replacement of hydroxyl groups) are described, much as for carbohydrates, by prefixes, as in **10** and **11**.

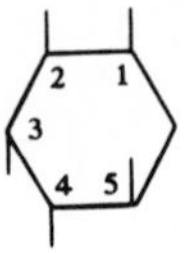

9 1D-1,2,5/3,4-Cyclohexanepentol (IUPAC/IUB)
2-Deoxy-D-*allo*-inositol (*CA*) after renumbering as the inositol)

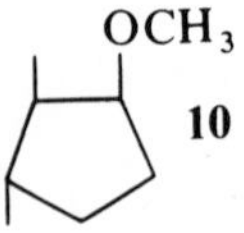

OCH$_3$

10 1D-1-*O*-Methyl-1,2/3-cyclopentanetriol (IUPAC/IUB)
[1*S*(1α,2β,3β)]-3-Methoxy-1,2-cyclopentanediol (*CA*)

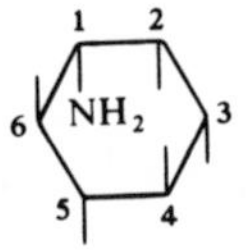

11 1D-1-Amino-1-deoxy-*myo*-inositol
1-Amino-1-deoxy-D-*myo*-inositol (*CA*)

Lipids

Recent recommendations for naming lipids[6] do not define the term, but they deal, as we do here, only with fats and related derivatives of glycerol. The most familiar lipids, esters of fatty acids and glycerol, are best named like esters of carbohydrates[3], i.e., by citing the acyl group(s) as prefixes to the parent name glycerol:

$$n\text{-}C_{17}H_{35}CO_2CH_2CH(O_2CC_{17}H_{35}\text{-}n)CH_2O_2CC_{17}H_{35}\text{-}n$$

tristearoylglycerol *or* glycerol tristearate

(Old names of the triglyceride and tristearin type should be abandoned.) A special constituent of lipids, 2*S*,3*R*-2-amino-1,3-octadecanediol **12**, is defined as sphinganine, and related compounds are named as derivatives thereof.

Lipids containing phosphoric acid as ester, i.e., phospholipids[7], are named systematically: 1,2-distearoyl-*sn*-glycerol 3-phosphate **13** (for

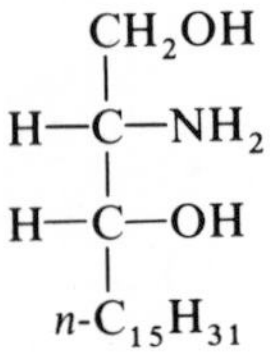

CH$_2$OH
|
H—C—NH$_2$
|
H—C—OH
|
n-C$_{15}$H$_{31}$

12

CH$_2$O$_2$CC$_{17}$H$_{35}$-n
|
n-C$_{17}$H$_{35}$CO$_2$CH
|
CH$_2$OPO$_3$H$_2$

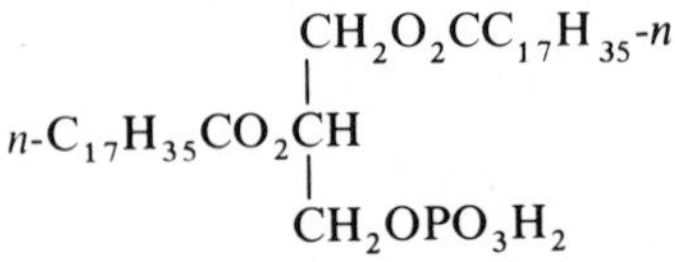

13

significance of the stereochemical prefix 'sn-' see p. 148). If such an acid esterifies, say, another glycerol molecule, the name is formed with a 'phospho-' infix, as in **14**, 1,2-distearoyl-*sn*-glycerol-3-phospho-3-*sn*-glycerol.

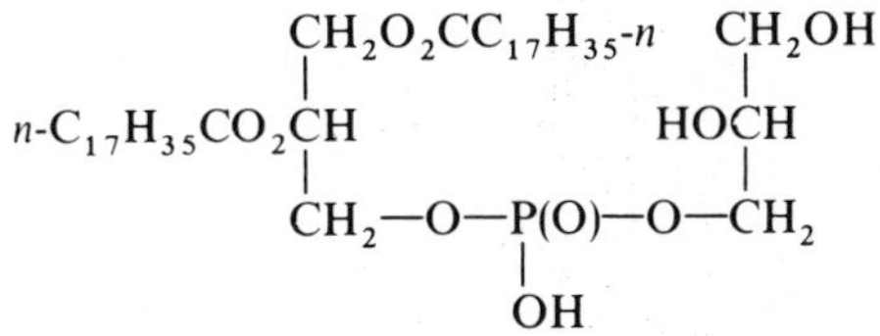

$$\underset{n\text{-}C_{17}H_{35}CO_2CH}{\overset{CH_2O_2CC_{17}H_{35}\text{-}n \quad CH_2OH}{|}}$$

$$\overset{HOCH}{\underset{CH_2-O-P(O)-O-CH_2}{|}}$$

$$\underset{OH}{|}$$

14

While systematic names are needed for individual lipids, generic ones less connected to structure often suffice. We merely mention phosphatidic acids, plasmenic acids, sphingolipids, and ceramides here; for structures and derivative names, the rules should be consulted.

Steroids

PARENT STRUCTURES AND STEREOCHEMISTRY

Steroids are cyclopenta[a] phenanthrene derivatives having the skeleton, numbering, and stereochemistry as shown in **15**. This general orientation

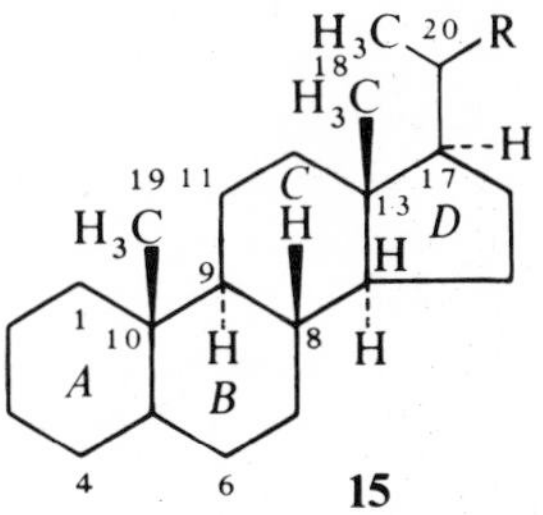

15

of the ring system is preferred, conforming to directions on p. 78. The individual rings are labelled *A* to *D*, as shown, for ease of reference.

The stereochemistry displayed (cf. p. 145) is that of common steroids. The IUPAC/IUB rules[8] permit specifying configuration at C-20 by a special use of *a* and *β* prefixes, but it is preferable to employ the R/S system here as for higher-numbered positions. Semisystematic names (for stereoparents) are then used by both IUPAC/IUB and CA[4c]:

> Gonane: lacks the side chains at C-17, C-18, and C-19.
>
> Estrane: has the C-18 methyl group but not the other side chains.
>
> Androstane: has the C-18 and C-19 methyl groups but no side chain at C-17.
>
> Pregnane: R = H (17β)
>
> Cholane: R = CH$_2$CH$_2$CH$_3$ (17β, 20R)
>
> Cholestane: R = CH$_2$CH$_2$CH$_2$CH(CH$_3$)$_2$ (17β, 20R)
>
> Ergostane: R = CH$_2$CH$_2$CH(CH$_3$)CH(CH$_3$)$_2$ (17β, 20R, 24S)
>
> Stigmastane: R = CH$_2$CH$_2$CH(C$_2$H$_5$)CH(CH$_3$)$_2$ (17β, 20R, 24R)

The stereochemistry specified is of course implied by the names, but that at C-5 is cited as 5α or 5β if known.

MODIFICATIONS

Unsaturation and substitution are described as usual in organic nomenclature.

A number of special structural prefixes are common in steroid names, and their use may be extended to other natural products[1]. These are only briefly noted here; for full directions on their use, the IUPAC/IUB or CA rules should be consulted.

'Cyclo-' signifies the formation of an additional ring by a direct link between carbon atoms of the parent structure, as in **16**.

'Nor-' denotes the elimination of an acyclic carbon atom, as in **17**, or of one from a ring, as in **18**.

'Homo-', the opposite of 'nor-', describes the addition of a carbon atom to expand a ring, again with a ring identifier; homo- and nor- may both be used in the same name, as in **19**.

'Seco-' (from Latin *secare* = to cut) signifies ring scission with addition of hydrogen at each of the positions previously linked, as in **20**.

'*Friedo-*' (shift of an angular methyl group) and '*A-neo-*' (contraction of ring A and other change), are less commonly used. The prefix '*abeo-*' (Latin = 'I go away'), meaning migration of a bond, is also not well known, but since IUPAC/IUB recommend that it replace the other two, it is illustrated in **21**.

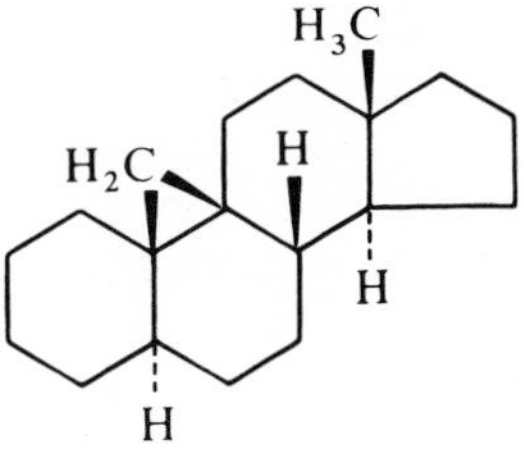

16

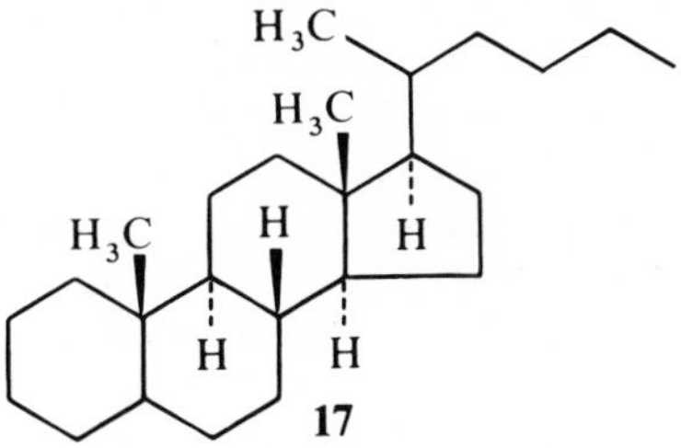

17

26,27-Dinorcholestane

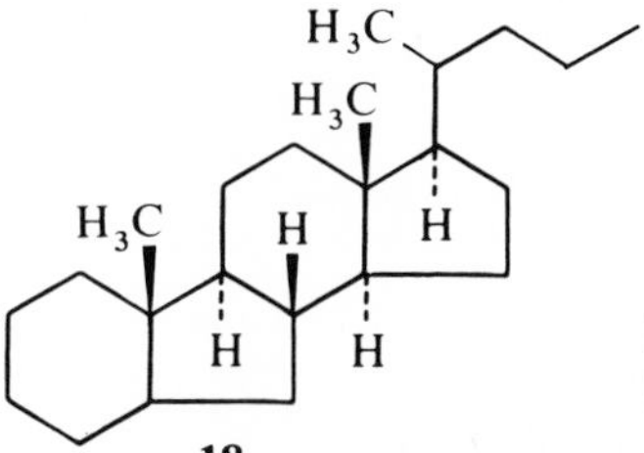

18

B-Norcholane

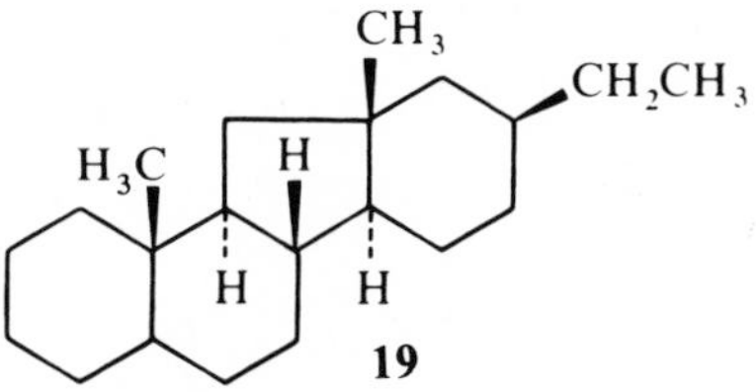

19

D-(17a)-Homo-*C*-norpregnane

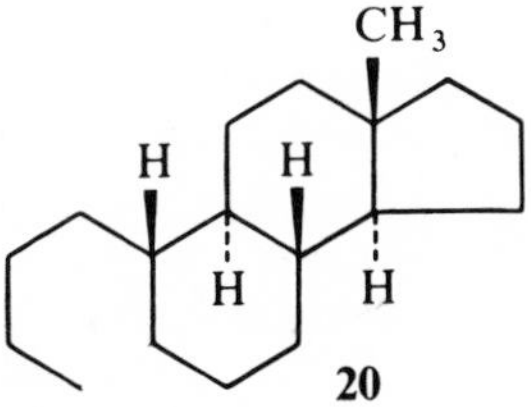

20

4,5-*Seco*estrane

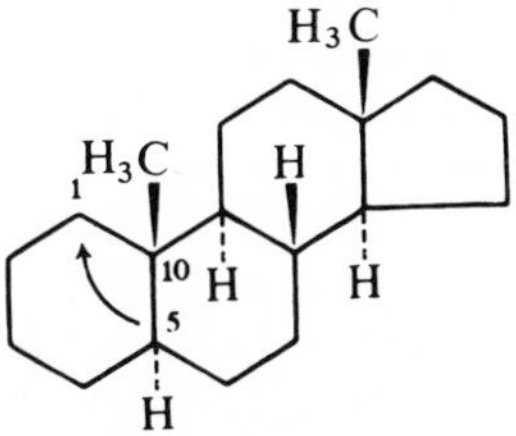

5α-Androstane

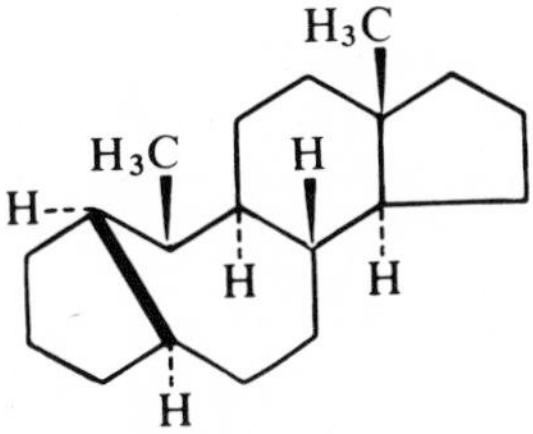

5(10→1) *abeo*-1α(*H*), 5α-Androstane

21

Terpenes

Acyclic terpenes ($C_{10}H_{16}$) are named systematically, but semisystematic names for parent cyclic monoterpenes are authorized by IUPAC[9]. The structures have fixed numbering as shown in *Table 8.1*, but the names imply no stereochemistry. This is of course readily described by the *R/S* system. Recent IUPAC tentative rules[1] having restricted the

Table 8.1 IUPAC PARENT NAMES FOR MONOTERPENES

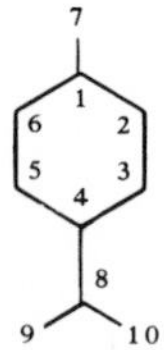

Menthane
(3 forms:
p- shown)

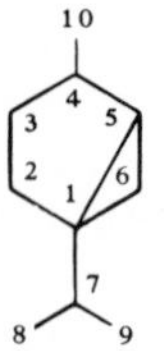

Thujane

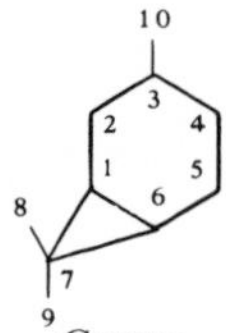

Carane

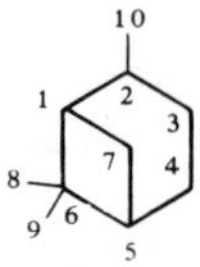

Pinane

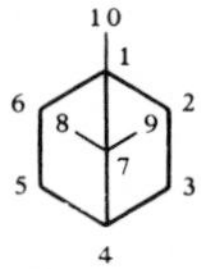

Bornane

meaning of the prefix 'nor-', the previously approved parent names of that type (norcarane, norpinene, norbornene) for the fully demethylated forms are not in good standing. The parent names may of course be modified by the usual substitutive and other prefixes, as in 7-chloro-*m*-menthane, or converted into radical names, as in 3-pinanyl. No similar names for sesquiterpenes ($C_{15}H_{24}$), diterpenes, and triterpenes have been adopted by IUPAC, although a number are in use. *Chemical Abstracts*[4d] names most mono-, sesqui-, and di-terpenes fully systematically, i.e., without special parent names, but does use stereoparents for tetracyclic terpenes and some others of complex stereochemistry.

Carotenoids are hydrocarbons (carotenes) and their oxygenated derivatives (xanthophylls, etc.) that may be considered derived from **22**, carotene[4d,10]. The terminal portions, C-1 to C-9 and C-1' to C-9',

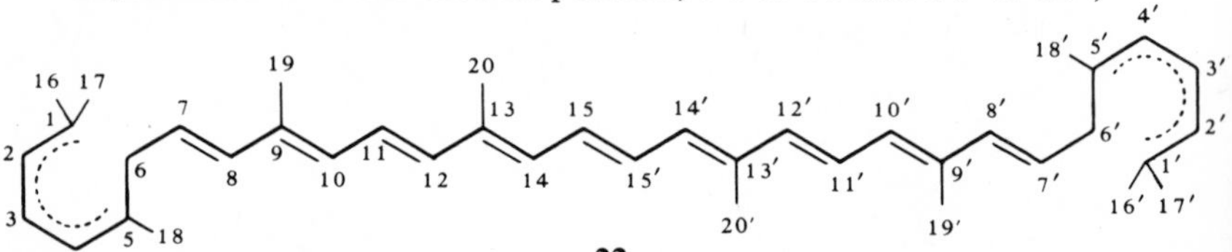

22

may have open-chain or 5- or 6-membered ring structures, and lower-case Greek letters (α, β, etc.) are prescribed to distinguish them. The prefix 'retro-', with two locants, is used to indicate a shift, by one position, of all single and double bonds of the conjugated system delimited by the locants. Except for these novelties, the affixes used to cite substituents and other modifications, as well as to describe stereochemistry, are those used in other organic nomenclature, especially that of steroids. While trivial names are not banned, they should not be used without concurrent citation of the carotene-based names.

Amino Acids and Peptides

AMINO ACIDS

The 20 α-amino acids under direct genetic control have trivial names, such as glycine and cystine, that have long been in use and are author-ized by IUPAC/IUB[11]; CA^{4e} uses 30 such stereoparent names. These stereoparent names are not modified by CA to describe the correspond-ing simple salts, esters, and amides, which are named systematically: $CH_3 CH(NH_2)CONH_2$, 2-aminopropanamide. The IUPAC/IUB rules name these derivatives by applying the usual suffixes, although this requires changing a terminal '-e' (or '-an', in tryptophan) instead of the usual '-ic acid': sodium alaninamide. Even aldehydes, $RCH(NH_2)CHO$, and primary alcohols, $RCH(NH_2)CH_2OH$, corresponding to the 20 amino acids may be named according to IUPAC/IUB by similar use of the endings '-al' and '-ol', as in $CH_3 CH(NH_2)CHO$, alaninal, and $HOCH_2 CH(NH_2)CH_2 OH$, serinol. However, the ready availability of more systematic names makes this an undesirable practice for chemists.

Configurations of α-amino acids are best expressed by the R/S system; in most natural ones, the α-carbon atom is S. In an older system the letters D and L referring to serine as standard was used (*see* p. 144).

PEPTIDES

Peptides are polymers of the type $H_2 N(CHRCONH)_n CHRCO_2 H$; in a dipeptide n is 1, in a tripeptide n is 2, etc. Peptides are named as deriva-tives of the amino acid retaining the carboxyl group (the carboxyl-terminal acid) by the use of one or more acyl names as prefixes, as in:

glycyl-(S)-alanine $\quad H_2 NCH_2 CONHCH(CH_3)CO_2 H$

(S)-alanyl-(S)-alanyl-(S)-threonine
$H_2 NCH(CH_3)CONHCH(CH_3)CONHCH(CHOHCH_3)CO_2 H$

(These formulas do not show the stereochemistry.) Proteins for which amino acid sequence is unknown of course have only trivial names.

Homopolymers of amino acids may also be named by placing a multiplying prefix, such as di- or in general poly-, before the name of the monomer, e.g., tetraglycine. A more general and structure-descriptive system is applicable, that based on simplest repeating unit (*see* p. 177), but it produces names of greater complexity.

ABBREVIATIONS

The three-letter abbreviations of trivial amino acid names authorized by IUPAC/IUB[12] — Gly, Lys, Tyr, etc. — representing either the whole molecule or a derived radical, are useful, especially for describing amino acid sequence in peptides. A single-letter system has also been approved[13] but is less often seen. There are rules for the nomenclature, often employing abbreviations, of synthetic polypeptides[14], synthetic modifications of natural peptides[15], peptide hormones[16], and iron—sulfur proteins[17].

ENZYMES

The special class of catalytic proteins called enzymes have long been designated by the names of their substrates, nearly always modified by the ending '-ase' and sometimes with words describing the reaction involved. Since this system is unwieldy and sometimes ambiguous, it is better to use the numerical classification approved by IUPAC/IUB[18]; e.g., catalase is E.C. (Enzyme Commission) 1.11.1.6. There are special rules[19] for multiple forms of enzymes, including isoenzymes.

Nucleic Acids, Nucleosides, and Nucleotides

The two well known types of nucleic acid are ribonucleic acid(s), RNA, and deoxyribonucleic acid(s), DNA. These are polymers made up of heterocyclic bases attached to a chain of alternating phosphoric acid and ribose or deoxyribose units, to which they may be hydrolysed. The commonest bases — adenine, guanine, xanthine, hypoxanthine, thymine, cytosine, and uracil — are so named by IUPAC, but *CA* uses only the systematic purine—pyrimidine names. The glycosyl bases are called nucleosides, and the ones most often met are named by both agencies by altering the base name to end in -osine or -idine, as in adenosine **23** and thymidine **24**. Nucleotides are phosphoric esters of nucleosides, and

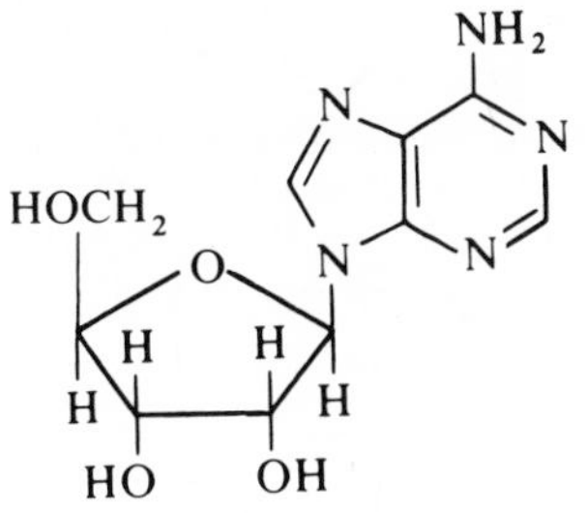

23

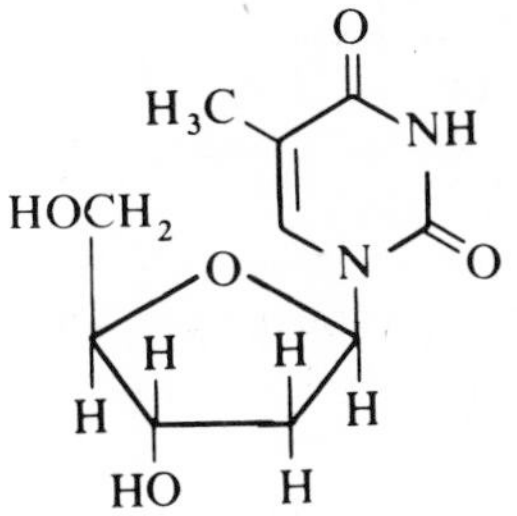

24

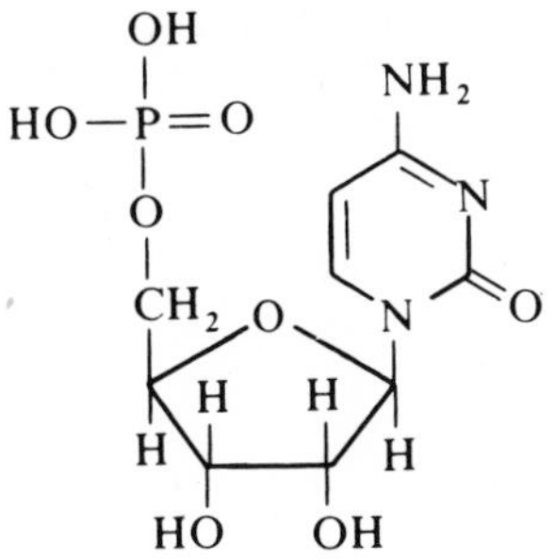

25

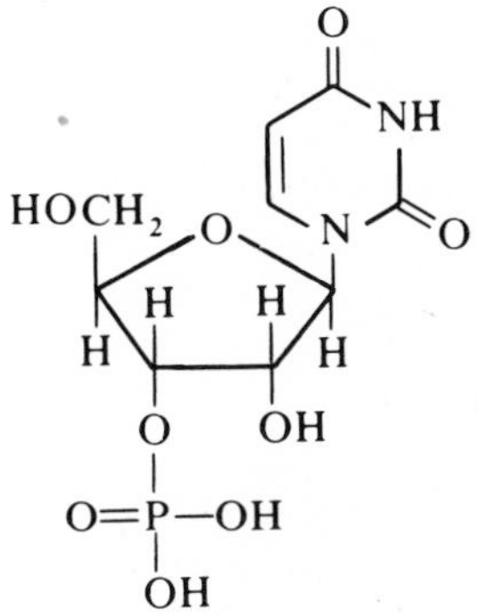

26

may be so named, a locant for the nucleoside portion usually being needed, as in cytidine 5'-phosphate **25**. However, the common ones are named as '-ylic acid' derivatives of nucleosides: 3'-uridylic acid **26**.

Abbreviations and symbols for representing these compounds have been recommended by IUPAC/IUB[20].

Alkaloids

Alkaloid nomenclature has not been systematized by IUPAC/IUB. *Chemical Abstracts*[4f] names very simple ones systematically, e.g., nicotine is indexed at (S)-3-(1-methyl-2-pyrrolidinyl)pyridine. Some 40 Class B stereoparents (*see* p. 151) were used as index names from 1972 to 1976; considerably more Class C stereoparents were likewise in use. Special structural and configurational prefixes [nor, homo, seco, retro, *enantio* (= *ent*)] have their usual meanings, except that subtraction of a methyl group from a nitrogen atom is denoted by '-demethyl-', not 'nor-'.

Porphyrins

Porphyrins contain the tetrapyrrole macrocyclic ring shown in **27**, and corrins **28** have the similar structure lacking one bridging carbon atom. The former have long been known, as in hemins (hemoglobins)* and chlorins (chlorophylls); the corrins are represented by B_{12} vitamins

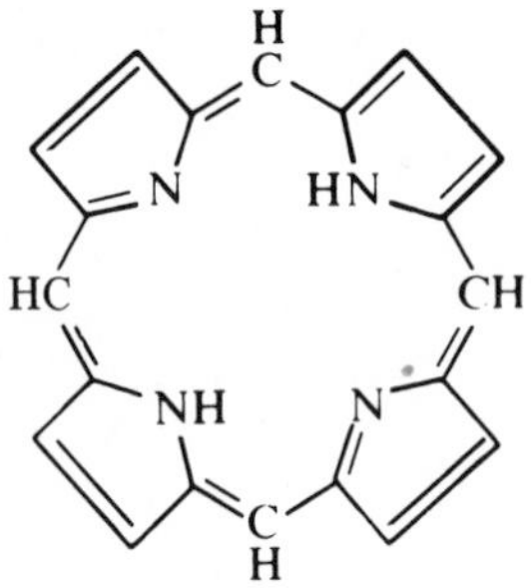

27

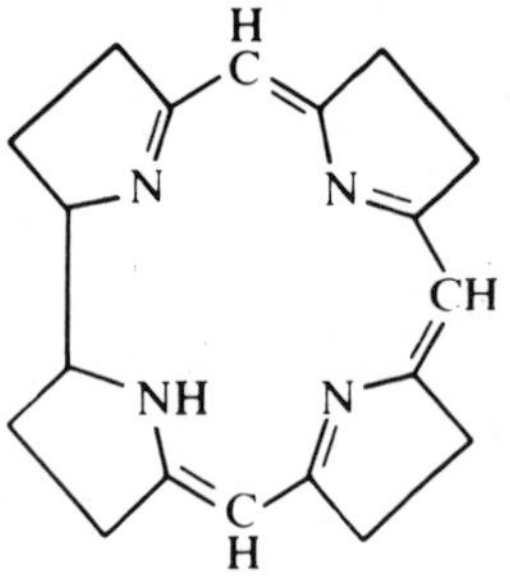

28

(*see* p. 165). *Chemical Abstracts* indexes porphyrins as derivatives of $21H,23H$-porphine, and names specific corrins, as IUPAC/IUB do, as cobamides and their relatives. Some abbreviations for use in corrinoid naming are authorized by IUPAC/IUB[21].

Vitamins

The vitamins are structurally diverse. Their biological activities are still best referred to by the old letter names, e.g., vitamin A activity. Individual members of the families are named as follows by *CA*[4g]; IUPAC/IUB recommendations are also given when they exist. Since no new principles of nomenclature are involved, no structures are shown here; they are readily accessible in *CA*.

A: both *CA* and IUPAC/IUB[22] recommend the names retinol, retinal, retinoic acid, etc.

B_1: named systematically (not as thiamine).

B_2: named as riboflavin and derivatives.

B_3, B_5: named systematically.

---

*The spellings haemins, hæmins, haemoglobin, hæmoglobin, and also haem, hæm, and heme are also found.

B$_6$: named systematically as pyridines. IUPAC/IUB[23] authorize semisystematic names based on the stem 'pyridox-': pyridoxine, pyridoxal, pyridoxamide, etc.

B$_{12}$: both *CA* and IUPAC/IUB[21] use vitamin B$_{12}$ as a name for a specific compound, but describe derivatives as arising from cobyrinic acid, etc. (cf. p. 164).

C: both *CA* and IUPAC/IUB[22] name vitamin C and its derivatives on the basis of ascorbic acid.

D: named systematically as steroids by *CA*, but IUPAC/ IUB[22] permit use of the name calciferol.

E: named systematically by *CA*. IUPAC/IUB[24] authorize the stereoparent name tocol, and even the old tocopherol names if *R/S* stereochemical descriptors are added.

K: named systematically by *CA*. According to IUPAC/IUB[25] special names ending in '-quinone', such as ubiquinone and phylloquinone, may be used either for groups of compounds or for individual ones. In the latter case a suffix numeral is used to express the number of intact isoprene units in the side chain attached to the quinone ring.

REFERENCES

1. *Nomenclature of Organic Chemistry: Section F: Natural Products and Related Compounds (Provisional)*, IUPAC Information Bull., Appendix No. 53 (1976)
2. *Compendium of Biochemical Nomenclature and Related Documents* (IUB), 3rd Ed., The Biochemical Society, P.O. Box 32, Colchester CO2 8HP, Essex (1978)
3. IUPAC/IUB *Carbohydrate Nomenclature 1 (Tentative Rules)*, 1969; *Biochem. J.*, **125**, 675 (1971); *Biochemistry*, **10**, 3983 (1971)
4. *Chemical Abstracts Vol. 76 Index Guide* (1972) or *Ninth Collective Index, Vols. 76–85 (1972–1976) Index Guide*, (*a*) para. 208 (*b*) para. 209 (*c*) para. 211 (*d*) para. 212 (*e*) para. 206 (*f*) para. 204 (*g*) para. 224
5. IUPAC/IUB, *Nomenclature of Cyclitols (Recommendations 1973)*, *Pure Appl. Chem.*, **37**, 283 (1974); *Biochem. J.*, **153**, 23 (1976)
6. IUPAC/IUB, *The Nomenclature of Lipids (Recommendations 1976)*, Hoppe-Seylers Z. Physiol. Chem., **358**, 617 (1977); *Mol. Cell. Biochem.*, **17**, 157 (1977)
7. IUPAC/IUB, *Nomenclature of Phosphorus-Containing Compounds of Biological Importance (Recommendations 1976)*, Proc. Natl. Acad. Sci. U.S.A., **74**, 2222 (1977)
8. IUPAC/IUB, *Nomenclature of Steroids (Rules 1971)*, *Pure Appl. Chem.*, **31**, 283 (1972)
9. *IUPAC Nomenclature of Organic Chemistry. Definitive Rules for: Section A. Hydrocarbons; Section B. Fundamental Heterocyclic Systems; Section C. Characteristic Groups Containing Carbon, Hydrogen, Oxygen, Nitrogen, Halogen, Sulfur, Selenium and/or Tellurium*, 1969. A, B 3rd Ed.; C 2nd Ed.,

Butterworths, London (1971) (now available from Pergamon Press, Oxford), p. 49

10. IUPAC/IUB, *Nomenclature of Carotenoids (Rules Approved 1974), Pure Appl. Chem.*, **41**, 405 (1975); *Eur. J. Biochem.*, **25**, 397 (1972) and **57**, 317 (1975)
11. IUPAC and IUPAC/IUB, *Nomenclature of α-Amino Acids (Recommendations, 1974)*, *Biochemistry*, **14**, 445 (1975); *Eur. J. Biochem.*, **53**, 1 (1975)
12. IUPAC/IUB, *Symbols for Amino-Acid Derivatives and Peptides, Recommendations (1971)*, *Pure Appl. Chem.*, **40**, 315 (1974); *J. Biol. Chem.*, **247**, 977 (1972)
13. IUPAC/IUB, *A One-Letter Notation for Amino Acid Sequences (Definitive Rules); Pure Appl. Chem.*, **31**, 639 (1972)
14. IUPAC/IUB, *Abbreviated Nomenclature of Synthetic Polypeptides (Polymerized Amino Acids); Pure Appl. Chem.*, **33**, 437 (1973)
15. IUPAC/IUB, *Definitive Rules for Naming Synthetic Modifications of Natural Peptides; Pure Appl. Chem.*, **31**, 647 (1972)
16. *Nomenclature of Peptide Hormones; IUPAC Information Bull.*, Appendix No. 48 (1975)
17. *Nomenclature of Iron–Sulfur Proteins; IUPAC Information Bull.*, Appendix No. 32 (1973)
18. IUPAC/IUB, *Enzyme Nomenclature (Recommendations 1972)*, Elsevier, Amsterdam, London, and New York, 1973; Supplement 1, *Biochim. Biophys. Acta*, **429**, 1 (1976)
19. *Multiple Forms of Enzymes; IUPAC Information Bull.*, Appendix No. 68
20. IUPAC/IUB, *Abbreviations and Symbols for Nucleic Acids, Polynucleotides, and their Constituents (Rules Approved 1974); Pure Appl. Chem.*, **40**, 277 (1974); *Eur. J. Biochem.*, **15**, 203 (1970)
21. IUPAC/IUB, *Nomenclature of Corrinoids (Rules Approved 1975); Pure Appl. Chem.*, **48**, 495 (1976)
22. IUPAC/IUB, *Trivial Names of Miscellaneous Compounds of Importance in Biochemistry, Biochim. Biophys. Acta*, **107**, 1 (1965); *J. Biol. Chem.*, **241**, 2987 (1966)
23. IUPAC/IUB, *Definitive Nomenclature for Vitamins B-6 and Related Compounds; Pure Appl. Chem.*, **33**, 445 (1973)
24. IUPAC/IUB, *Nomenclature of Tocopherols and Related Compounds, Recommendations (1973)*, *Arch. Biochem. Biophys.*, **165**, 6 (1974); *Eur. J. Biochem.*, **46**, 217 (1974)
25. IUPAC/IUB, *Nomenclature of Quinones with Isoprenoid Side-chains; Pure Appl. Chem.*, **38**, 439 (1974); *Biochem. J.*, **147**, 15 (1975)

9

Miscellaneous Nomenclature

Organometallic Compounds

The rules of nomenclature thus far presented have not systematically covered compounds in which organic groups are linked through carbon to atoms other than carbon, hydrogen, nitrogen, halogens, and chalcogens. While it is questionable to classify compounds containing C–B, C–Si, and C–P bonds as organometallic (since boron, silicon, and phosphorus do not behave much like metals), they are discussed here. Some organometallic compounds have already been mentioned as π-complexes (*see* p. 29).

In this area, where inorganic chemistry and organic chemistry intersect, chemists in these two fields have mostly sought to maintain their own systems of nomenclature, which are based on description of structure in inorganic, but of function (when possible) in organic chemistry. This has led to sanction of even more alternative names than usual in the tentative IUPAC organic nomenclature recommendations, Section D^1, on which this discussion is based. These Rules, which were issued jointly by both the Commissions immediately concerned, should be consulted for more detail than can be given here. The naming practices of *CA* are also summarized in the following discussion.

NAMES OF COORDINATION COMPOUNDS

As noted on p. 17 and illustrated thereafter, the extended coordination principle is broadly applicable. In organometallic compounds the metal is naturally regarded as central; we ignore the uncommon organometallics that contain two metal atoms. This produces names such as:

tetraethyllead	$(C_2H_5)_4Pb$
oxotriphenylphosphorus	$(C_6H_5)_3PO$
trimethyl(pyridine)boron	$(C_5H_5N)B(CH_3)_3$
hydroxo(phenyl)mercury	C_6H_5HgOH
(2-formylethyl)tripropylgermanium	$(n\text{-}C_3H_7)_3GeCH_2CH_2CHO$

Polymeric species are often named as if monomeric, e.g., $(CH_3Li)_n$, methyllithium. Coordination names are used extensively by *CA*, but usually not for, e.g., the elements whose hydrides are listed in *Table 9.1*; hexacoordinate tin compounds are named by coordination principles.

SUBSTITUTIVE NAMES

As described in Chapter 4, constructing a substitutive name involves choosing a parent compound and citing replacement of hydrogen therein, one substituent (the principal function) being cited as suffix and the others as prefixes. In organometallic compounds it is often possible to choose either a metal hydride or an organic compound as parent, although very different names result. In *CA* names the organic part is chosen as parent if it contains any functional group cited as suffix.

Table 9.1 PARENT HYDRIDE NAMES

Arsine	AsH_3	Plumbane	PbH_4
Arsorane	AsH_5	Silane	SiH_4
Bismuthine	BiH_3	Stannane	SnH_4
Borane	BH_3	Stibine	SbH_3
Germane	GeH_4	Sulfane	SH_2
Phosphine	PH_3	(used only in disulfane,	
Phosphorane	PH_5	etc.; *see* p. 124)	

Simple hydrides recognized as parents by IUPAC are shown in *Table 9.1*. Then, e.g., $(C_6H_5)_3P$ is triphenylphosphine, $(CH_3)_4Si$ tetramethylsilane, $(CH_3)_3AsO$ trimethyloxoarsorane, and $(CH_3O)_3GeCH_2CO_2CH_3$ trimethoxy(methoxycarbonylmethyl)germane [but methyl (trimethoxygermyl)acetate in *CA*].

The suffixes and prefixes used in organic nomenclature to express functional groups serve for naming compounds in which these are attached to silicon. According to IUPAC, they may be used in either form, e.g., $(CH_3)_3SiOH$, hydroxytrimethylsilane or trimethylsilanol, but *CA* uses only the latter kind of name for silicon compounds. For other elements, functional suffixes must be used with caution. Thus $(CH_3)_2POH$ may be named as dimethylphosphinous acid (*see* p. 170), but R_2PCO_2H is acceptably called a dialkylphosphinecarboxylic acid. Substituents in borane are cited by IUPAC only as prefixes, as in $(CH_3)_2BNH_2$, aminodimethylborane; $CH_3B(OH)_2$, dihydroxy(methyl)-borane; and $(CH_3)_2BOH$, hydroxydimethylborane. The *CA* names for these are respectively dimethylboranamine, methylboronic acid, and dimethylborinic acid. Even $(HO)_3B$ may be called trihydroxyborane by IUPAC rules, but the name boric acid or orthoboric acid is equally acceptable and certainly more familiar.

When metallic radicals are substituents in parent organic compounds, they are not usually regarded as functional and are therefore cited as prefixes. *Table 9.2* shows some prefix names corresponding to the

Table 9.2 ORGANOMETALLIC RADICAL AND PREFIX NAMES (IUPAC)

Arsino	$-AsH_2$	Phosphinediyl	$>PH$
Arsinediyl	$>AsH$	Phosphoranyl	$-PH_4$
Arsoranyl	$-AsH_4$	Plumbyl	$-PbH_3$
Bismuthino	$-BiH_2$	Silyl	$-SiH_3$
Boryl	$-BH_2$	Silanediyl	$>SiH_2$
Boranediyl	$>BH$	Stannyl	$-SnH_3$
Germyl	$-GeH_3$	Stibino	$-SbH_2$
Phosphino	$-PH_2$		

hydrides of *Table 9.1*. The current *CA* prefix names are the same as those in *Table 9.2* except for the bivalent ones: $>AsH$ arsinidene, $>BH$ borylene, $>PH$ phosphinidene, and $>SiH_2$ silylene. For all other elements (except C, H, O, halogens, chalcogens and N) the radical names are formed by changing the element name so as to end in '-io', e.g., thallio. The names of inorganic and organic ligands are prefixed to these as appropriate to make group names such as diethylaluminio $-Al(C_2H_5)_2$, chloromercurio $-HgCl$, and hydridoberyllio $-BeH$. The '-io' names may also be used instead of those in *Table 9.2*, and for multivalent radicals as well. Thus we have names such as (dimethylarsino)acetic acid or (dimethylarsenio)acetic acid for $(CH_3)_2AsCH_2CO_2H$, and 4,4'-(diethylstannio)diphenol, $(C_2H_5)_2Sn(C_6H_4OH\text{-}p)_2$. Univalent metals linked to carbon may be cited by -io- infixes, as in 1-lithiobutane, $LiCH_2CH_2CH_2CH_3$. The status of 'deuterio' and 'tritio' as substitutive prefixes is unclear; they are authorized by IUPAC rules for inorganic compounds[2a], but not mentioned by the more recent provisional ones for organic compounds[3].

Organometallic anions not having O, S, or N linked to the metal atom are named like the radicals, but with the -io ending changed to -ate: LiCu(CH$_3$)$_3$, lithium trimethylcuprate; [(C$_6$H$_5$)$_3$Pb]$^-$, triphenylplumbate. The oxidation number of the metal is shown if necessary by a Stock number, or the charge on the anion by a Ewens–Bassett number (*see* p. 16), as in [(C$_6$H$_5$)$_4$P]$^-$, tetraphenylphosphate(III) or tetraphenylphosphate(1−) (the standard name phosphate is a contraction of phosphorate specified by the rule); [(CH$_3$)$_2$SiF$_4$]$^{2-}$, tetrafluorodimethylsilicate(2−). The old-fashioned '-ide' ending for describing carbanions, e.g., in NaCH$_3$, sodium methanide, is authorized by IUPAC Section C rules[5] but not by those of Section D[1]. The latter, however, do specify names such as potassium diethylphosphide, KP(C$_2$H$_5$)$_2$. Acetylide names such as (mono)sodium acetylide, NaC≡CH, are still used by both IUPAC and *CA*, although ethynylsodium is certainly also an acceptable name.

RADICOFUNCTIONAL NAMES

Organometallic compounds containing anionic components may be named, besides the methods already mentioned, by citing the organic group(s) plus the metal as one word and the anionic component(s) as other(s), e.g., methyltin chloride dihydride, phenylmagnesium bromide. Such multiple-word names should be used rather than the old 'acid' names for R$_x$M(O)$_y$(OH)$_z$, when M = Ge, Sn, Pb, Sb, or Te, since these compounds are both polymeric and amphoteric. Thus C$_6$H$_5$Sb(O)(OH)$_2$ is to be called phenylantimony dihydroxide oxide (cf. p. 26) and not benzenestibonic acid.

Organometallic cations may be named similarly, by prefixing the proper ligand names to the name of the metal and giving the Stock number or Ewens–Bassett number, e.g., [(C$_2$H$_5$)$_3$Sn]$^+$, triethyltin(IV) or triethyltin (1−). The '-onium' system (*see* pp. 23 and 103) is sometimes applicable, as in [(HOCH$_2$)$_4$P]$^+$ Cl$^-$, tetrakis(hydroxymethyl)phosphonium chloride.

Compounds R$_x$M(O)$_y$(OH)$_z$ when M is P or As, and their analogues and derivatives, have special names based on the parent oxoacid names shown in *Table 9.3* (cf. p. 19). Replacement of the hydrogen on phosphorus or arsenic by organic groups is denoted by prefixes as

Table 9.3 PARENT OXOACIDS OF PHOSPHORUS AND ARSENIC

H$_2$POH	phosphinous acid	H$_2$P(O)OH	phosphinic acid
H$_2$AsOH	arsinous acid	H$_2$As(O)OH	arsinic acid
HP(OH)$_2$	phosphonous acid	HP(O)(OH)$_2$	phosphonic acid
HAs(OH)$_2$	arsonous acid	HAs(O)(OH)$_2$	arsonic acid

usual, e.g., $[(C_6H_5)_2P-O]^- Na^+$ sodium diphenylphosphinite. Replacing an $-OH$, $=O$, or both is described by inserting a special name (an infix) just before the last syllable ('-ic', '-ate', '-ous', or '-ite') of the acid name. For a complete list of infixes the complete rules should be consulted, but they have the form '-chlorid-' or '-chlorido-', '-amid-' or '-amido-', '-thio-', etc. Thus are formed names such as

$CH_2=CHP(OCH_3)Cl$ methyl vinylphosphonochloridite
$1\text{-}C_{10}H_7As(O)(NH_2)OH$ 1-naphthylarsonamidic acid

However, if no $-OH$, $-OR$, or $-O-$metal group remains on P or As, the compound is named as an acid chloride, amide, etc.:

$C_2H_5AsCl_2$ ethylarsenous dichloride
$[(CH_3)_2CH]_2PNHCH_3$ diisopropylphosphinous methylamide

Other names for these, and for radicals corresponding to the acids of *Table 9.3*, may be found in the IUPAC text[1].

NAMES BASED ON 'a' TERMS

The 'a' terms are names derived from those of the elements by variously modifying them to end in 'a'; we have already encountered the commonest ones in *Table 5.5*. They represent atoms constituting parts of rings or chains.

In one application, one or more 'a' terms are prefixed to the name of an organic compound to indicate replacement of the carbon atom(s) specified (replacement nomenclature, *see* pp. 56, 98). As already noted, this system is rarely needed among acyclic compounds, but it is easily applied and occasionally advantageous for cyclic structures, as in 1, 2, and 3. In *CA* it is used only rarely, as for 2.

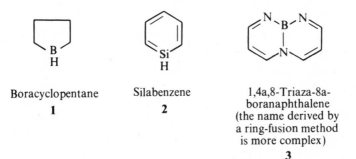

Boracyclopentane   Silabenzene   1,4a,8-Triaza-8a-
1   **2**   boranaphthalene
                                  (the name derived by
                                  a ring-fusion method
                                  is more complex)
                                  3

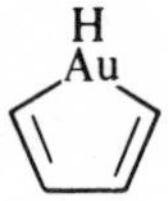

1H-Aurole
4

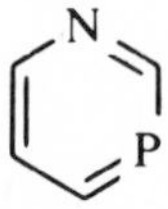

1,3-Azaphosphorine
5

However, the Extended Hantzsch–Widman system (*see* p. 90) works well and is preferred by *CA* for cyclic organometallic compounds such as **4** and **5**.

Finally, chains and rings containing regular patterns of atoms other than carbon need consideration. Homogeneous chains are named with a multiplying prefix, the appropriate 'a' term, and the ending '-ne', e.g., $H_3Si–SiH_2–SiH_3$, trisilane (cf. p. 14). In such names only, the 'a' terms for sulfur, selenium, and tellurium are 'sulfa', 'sela', and 'tella' (as suggested on p. 14); but *CA* does not use these. For elements of variable valence, it must be made clear which valence is operative. To this end *connecting number* is defined as the sum of sigma bonds, pi bonds, and units of charge (positive or negative) on an atom. This is cited when necessary by using the symbol λ and a following superscript. In this way $S\lambda^6$ means sulfur with connecting number 6, as in the sulfate ion. 'Normal' connecting numbers, which do not have to be cited, are defined as 1 for halogens, 2 for chalcogens and Group II metals, 3 for elements of the boron and nitrogen families, and 4 for C, Si, Ge, Sn, and Pb. Thus $N_2N–NH–NH_2$ is triazane, but $H_4P–PH_3–PH_4$ (hypothetical, to be sure) tri-λ^5-phosphane. In systems of alternating atoms ending in identical metal atoms, both kinds are cited, as in tetrasiloxane, $H_3SiOSiH_2OSiH_2OSiH_3$, and cyclotriborazane **6** (trivial

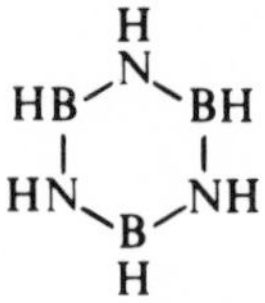

6

name, used by *CA*, borazine). However, when chains of repeating units are present, the prefix '*catena-*' is used, as for $H_2N–PH–NH–PH–NH–PH_2$, *catena*-tri(phosphazane). Branched chains, multiple bonds, rings, and radicals are described as in organic nomenclature.

Polyboranes pose special problems in a field growing so rapidly that specialists are already considering changes in rules approved in 1971. The presence of bridging hydrogen atoms and the variability of the connecting number of boron make unusual hydrides possible, and names must reflect the numbers of both boron and hydrogen atoms present, e.g., pentaborane(9), B_5H_9; pentaborane(11), B_5H_{11}. Polyboranes having closed three-dimensional structures may receive the prefix *closo*- and those with very nearly closed structures take the prefix *nido*- (Latin *nidus* = nest), along with numbering as illustrated in **7** and **8**.

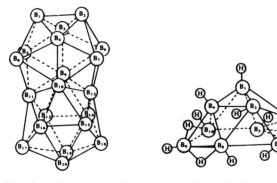

$B_{20}H_{16}$ *closo*-Icosaborane (16)

7

$B_{10}H_{14}$ *nido*-Decaborane (14)

8

Sometimes one or more boron atoms of a polyboron skeleton are replaced by other atoms, as in **9**, one of the isomeric forms of dicarba-*closo*-dodecaborane(12). (The *CA* name does not include the

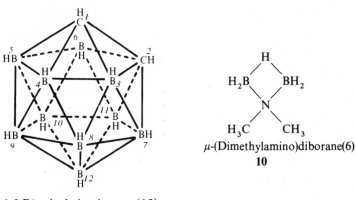

1,2-Dicarbadodecaborane (12)

9

μ-(Dimethylamino)diborane(6)

10

'*closo*'.) When replacement of a bridging *hydrogen* has occurred, as in **10**, the replacing group name carries the (bridging) symbol as prefix, if necessary with the locants of the atoms it links.

SUMMARY

As an aid in finding a name for a specific compound, the following condensation is intended to lead the reader to the specific paragraphs preceding:

(1)   Except for compounds of As, Bi, B, Ge, P, Pb, Si, Sn, Sb, or S(II), coordination names are always permissible.

(2)   Simple derivatives of these 'exceptional' elements are usually named from the hydride as parent.

(3)   When a functional group is present that can be cited as suffix, the organic part is chosen as parent in all *CA* names, with the organo-metallic unit as a substituent. This is, however, optional in IUPAC practice.

(4)   Special rules apply to silicon and boron and to oxoacids of P or As.

(5)   The radical prefix for an organometallic unit is the name of the metal with its ending changed to '-io-', except for the 'exceptional' elements named above; for the latter, radical names derived from the hydride are used.

(6)   Organometallic anions often have names ending in '-ate', although multiple-word radicofunctional names may be used in some cases.

(7)   Organometallic cations are also usually named by the radico-functional method.

(8)   Cyclic compounds containing a heteroatom in a ring or, less often, in a chain may be named by the 'oxa–aza' system, or, for rings, by the Hantzsch–Widman system. In chains of repeating units special nomenclature based on 'a' terms is available.

Isotopically Modified Compounds

Isotopically modified compounds are those in which the isotopic nuclide composition differs from that occurring in nature. In writing formulas or names for such compounds, the nuclide in excess is preferably represented by the usual atomic symbol with mass number shown by a left superscript numeral, e.g., ^{14}C (cf. p. 9). For hydrogen isotopes D and T are acceptable for otherwise unmodified compounds, and the suffixes *d* and *t* are used in *CA* names.

Two kinds of isotopically modified compound may be distinguished[3]; isotopically *substituted* compounds, in which essentially all molecules have only the indicated nuclide at each specified position, and isotopi-cally *labelled* compounds, which are mixtures of isotopically unmodified compounds, usually in overwhelming excess, with one or more analogous substituted compounds. In practice, radioactive nuclides are usually encountered in labelled compounds, and stable ones in substituted compounds.

The IUPAC name for an isotopically *substituted* organic compound is formed by inserting in parentheses the nuclide symbol(s), preceded by any necessary locant(s), immediately before the whole name or preferably before the name of the substituted part of the compound. As is done for substituted compounds in general, the number of atoms substituted is described by a subscript when polysubstitution is possible. This procedure yields names such as $(2\text{-}^{13}C)\text{-}1$-butene, or in biochemical usage $1\text{-}(2\text{-}^{13}C)$butene, for $CH_2 = {}^{13}CHCH_2CH_3$, and methyl $({}^2H_2)$acetate for $CH^2H_2CO_2CH_3$. The modified Boughton system used by CA^5 does not differentiate between substituted compounds and labelled compounds, and uses italic suffixes (again to the whole name or to that part modified) to tell location and nature of modifying nuclides. The two compounds just mentioned are thus called 1-butene-$2\text{-}^{13}C$ and methyl acetate-d_2.

A system like that of *CA* but specifying that the italicized suffix be enclosed in brackets is briefly described in IUPAC rules for inorganic compounds[2b]; this produces names such as phosphorus$[^{32}P]$ trichloride for $^{32}PCl_3$. It is to be hoped that this is soon abandoned in favour of either the IUPAC organic method or that of *CA*.

In representing *labelled* compounds, the nuclide symbol is placed in square brackets, instead of the parentheses used for substituted compounds. For *specifically labelled* ones, such as **11**, **12**, and **13**, in which the position(s) and number(s) of labelling nuclide(s) are known, names

$C[^3H_3]Cl$

Chloro$[^3H_3]$methane (IUPAC)
Chloromethane-t_3 (*CA*)

11

$CH_2 = C[^3H]CH_2CH_3$

2-$[^3H]$-1-Butene (IUPAC)
1-$[2\text{-}^3H]$Butene (biochemical usage)
1-Butene-$2\text{-}t$ (*CA*)

12

$CH_3CO_2[^{14}C]H_2CH_3$

$[1\text{-}^{14}C]$Ethyl acetate (IUPAC)
Ethyl-$1\text{-}^{14}C$ acetate (*CA*)

13

and formulas are almost like those for substituted compounds. Compounds multiply specifically labelled with the same or different nuclides are represented and named similarly.

When a compound is *selectively labelled*, the position(s) but not the number(s) of labelling nuclide(s) are known. The formula is then written without subscript(s) with the nuclide symbol(s) placed in front as usual, as in **14**. In naming *non-selectively labelled* compounds the square

$$[2,3\text{-}^{14}C]\,CH_3\,CH_2\,CH_2\,OH$$

14

[2,3-^{14}C]-1-Propanol or
[2,3-^{14}C] Propyl alcohol (IUPAC)
1-Propanol-*2,3-*^{14}*C (CA)*

brackets are used, but neither locants nor subscripts: [^{14}C] alanine. If *all* positions of a selectively labelled compound are labelled, the letter G may be inserted in the name, as in [G-^{3}H] pyridine; if all such positions are known to be labelled in the same isotopic ratio, the letter U is similarly used: [U-^{14}C] aniline.

In isotopically modified compounds the numbering of the unmodified compound is retained according to IUPAC rules; thus in (^{2}H)CH$_2$CH$_2$I, 1-iodo-2(^{2}H)-ethane, the iodine takes position 1. Of course in CH$_3$-[^{14}C]H$_2$−CH$_2$−CH$_3$ the labelling carbon atom is at position 2 and not 3. In *CA*, however, the modifying nuclide has priority over other substituents expressed as prefixes: (^{2}H)CH$_2$CH$_2$I is then named 2-iodoethane-*1-d*. Occasionally a modifying nuclide is located at a position that is not numbered; then an italicized prefix, etc., may be used as locant, as in C$_6$H$_5$CHOHO(=^{18}O)C$_6$H$_5$, [*carbonyl*-^{18}O] benzoin or benzoin-*carbonyl*-^{18}*O*.

Polymers

As in many other areas of chemical nomenclature, two kinds of name are applied to polymers. The structure-based approach names polymer molecules as made up of repeating units of atoms or groups, whereas source-based or process-based names, the older ones, merely add the prefix poly- to the monomer names. Sometimes the latter system is only conceptual since, e.g., poly(vinyl alcohol) is not actually made by polymerizing vinyl alcohol.

In spite of the advantages of the structure-based system, to which *CA* changed in 1972, we discuss the older names first; they are widely used, especially in technology and biochemistry, and are recognized by IUPAC[6]. Approved abbreviations of such names are also listed by IUPAC[7]. Thus we have, e.g., (−CH$_2$CH$_2$−)$_n$, polyethylene, PE; (−OCH$_2$CH$_2$−)$_n$, poly(ethylene oxide), PEO; and (−OCH$_2$CH$_2$O$_2$CC$_6$H$_4$CO−)$_n$ (*p*-), poly(ethylene terephthalate). Note the use of parentheses, which is needed to distinguish, e.g., poly(chloromethyl acrylate), which is polymerized chloromethyl

acrylate, from polychloromethyl acrylate, which must be either
$CH_2=CHCO_2CHCl_2$ or $CH_2=CHCO_2CCl_3$ or their mixture; then
polychloro(methyl acrylate) would be something else still. Copolymers
(as distinguished from homopolymers) may be named by interpolating
the italicized syllable '-*co*-' between the monomer names, as in
poly(acrylonitrile-*co*-styrene). In coordination compounds, linear
polymeric structures are designated by the prefix '*catena*' (*see* p. 33).

The structure of linear organic polymers is conveniently described
in terms of the smallest repeating unit. This system is now used by *CA*[8],
and has been approved by IUPAC[6], but IUPAC prefers the phrase
'constitutional repeating unit'. This unit may be composed of a single
bivalent group such as those on p. 73 or of a combination of these;
the name of the unit, placed in parentheses, is preceded by the prefix
'poly-'. Since it is possible to write and name unsymmetrical smallest
repeating units in two directions, it is necessary that the rules specify
the beginning and the direction of citation. This makes the rules too
complex to summarize here, but the results are illustrated by **15, 16**,
and **17**.

$(-CH_2-)_n$ $(-OCH_2CH_2-)_n$

poly(methylene)   poly(oxy-1,2-ethanediyl) (*CA*) or
                                               poly(oxyethylene) (IUPAC)

15   **16**

$(-NHCO\!-\!\bigcirc\!-\!NHCH_2CH_2CH_2CH_2-)_n$

Poly(iminocarbonyl-1,3-phenyleneimino-1,4-butanediyl) (*CA*)

Poly(iminocarbonyl-*m*-phenyleneiminotetramethylene) (IUPAC)
17

The IUPAC recommendations[6] extend to generic names for linear
polymers. Polymers consisting of only a few repeating units are oligomers.
When a polymer is made up of smallest repeating units in only one
sequence it is called regular; when substantial portions of a polymer
differ, such portions are blocks. Regular polymers may be tactic,
isotactic, or syndiotactic — often stereoregular — but for definitions of
these configuration-describing terms the rules must be consulted. There
are also special rules[9] on describing conformation of peptide chains.

Physicochemical Nomenclature

Probably the most useful publication of its kind is the *Manual of Symbols and Terminology for Physicochemical Quantities, 1973 Edition*[10]; it covers, of course, the SI system of units. It also includes an appendix on activities. Other appendices dealing with colloid and surface chemistry[11,12] and electrochemistry[13] have been issued. Further areas of official recommendations may be seen from the titles given in references[13-25].

Analytical Nomenclature

The IUPAC has issued more publications, mostly rather short, on analytical nomenclature than in any other field. Again the subjects of

Table 9.4 PREFIXES AND SYMBOLS FOR DECIMAL MULTIPLES OF UNITS

Multiple	Prefix	Symbol	Multiple	Prefix	Symbol
10^{-1}	deci	d	10^1	deca	da
10^{-2}	centi	c	10^2	hecto	h
10^{-3}	milli	m	10^3	kilo	k
10^{-6}	micro	μ	10^6	mega	M
10^{-9}	nano	n	10^9	giga	G
10^{-12}	pico	p	10^{12}	tera	T
10^{-15}	femto	f			
10^{-18}	atto	a			

these recommendations may be read from the list of references[26-50]. These also have been collected into one volume[51].

We also list in *Table 9.4* the prefixes to be used to construct decimal multiples of units.

Clinical Chemical Nomenclature

Two substantial sets of recommendations[52,53] on quantities and units in clinical chemistry have been issued.

REFERENCES

To save space, all publications in the following list for which no author is shown are to be understood as issued by one or more of the IUPAC Commissions. *Appendix* is used as an abbreviation for *IUPAC Information Bulletin, Appendices*

on *Provisional Nomenclature, Symbols, Terminology, and Conventions.* IUPAC
publications may be obtained (when available) from Pergamon Press, Oxford.

1. *Nomenclature of Organic Chemistry: Section D; Appendix* No. 31 (1973)
2. *IUPAC Nomenclature of Inorganic Chemistry, 2nd Ed., Definitive Rules 1970,*
   Butterworths, London (1971); (*a*) p. 10; (*b*) p. 12
3. *Nomenclature of Organic Chemistry: Section H. Isotopically Modified
   Compounds; Appendix* No. 62 (1977)
4. *Chemical Abstracts, Vol. 76, Index Guide* (1972) or *Ninth Collective Index,
   Vols. 76–85* (1972–1976), *Index Guide*, paragraph 220 (1977)
5. *IUPAC Nomenclature of Organic Chemistry. Definitive Rules for: Section A.
   Hydrocarbons; Section B. Fundamental Heterocyclic Systems; Section C.
   Characteristic Groups Containing Carbon, Hydrogen, Oxygen, Nitrogen,
   Halogen, Sulfur, Selenium and/or Tellurium,* 1969. A,B 3rd Ed.; C 2nd Ed.,
   Butterworths, London (1971)
6. *Nomenclature of Regular Single-Strand Organic Polymers (Rules Approved
   1975); Pure Appl. Chem.,* **48**, 375 (1976)
7. *List of Standard Abbreviations (Symbols) for Synthetic Polymers and
   Polymer Materials,* 1974; *Pure Appl. Chem.,* **40**, 475 (1974)
8. *Chemical Abstracts, Vol. 76 Index Guide,* (1972) or *Ninth Collective Index,
   Vols. 76–85* (1972–1976), *Index Guide*, paragraph 222 (1977)
9. IUPAC/IUB, *Abbreviations and Symbols for Description of Conformation of
   Peptide Chains; Pure Appl. Chem.,* **40**, 291 (1974)
10. *Manual of Symbols and Terminology for Physicochemical Quantities and
    Units,* 1973 Ed., Butterworths, London, 1975; now available from Pergamon
    Press, Oxford
11. *Definitions, Terminology and Symbols in Colloids and Surface Chemistry – I;
    Pure Appl. Chem.,* **31**, 577 (1972)
12. *Definitions, Terminology and Symbols in Colloid and Surface Chemistry – II;
    Pure Appl. Chem.,* **46**, 71 (1976)
13. *Electrochemical Nomenclature; Pure Appl. Chem.,* **37**, 499 (1974); *cf.
    Pure Appl. Chem.,* **45**, 131 (1976)
14. *A Guide to Procedures for the Publication of Thermodynamic Data; Pure
    Appl. Chem.,* **29**, 395 (1972)
15. *Recommendations for the Presentation of NMR Data for Publication in
    Chemical Journals – A. Conventions Relating to Proton Spectra; Pure Appl.
    Chem.,* **29**, 625 (1972); and *– B. Conventions Relating to Spectra from Nuclei
    Other than Protons; Pure Appl. Chem.,* **45**, 217 (1976)
16. *Recommendations for the Presentation of Raman Spectra for Cataloging and
    Documentation in Permanent Data Collections; Pure Appl. Chem.,* **36**, 275
    (1973)
17. *Nomenclature and Conventions for Reporting Mössbauer Spectroscopic Data;
    Pure Appl. Chem.,* **45**, 211 (1976)
18. *Nomenclature and Spectral Presentation in Electron Spectroscopy Resulting
    from Excitation by Photons; Pure Appl. Chem.,* **45**, 221 (1976)
19. *Recommendations for the Presentation of Infrared Absorption Spectra in
    Data Collections, A. Condensed Phases; Pure Appl. Chem.,* **50**, 231 (1978)
20. *Symbolism and Nomenclature for Mass Spectroscopy, Appendix* No. 51
    (1976); *cf. Pure Appl. Chem.,* **37**, 569 (1974)
21. *Definition and Symbolism of Molecular Force Constants, Appendix* No. 56
    (1976)
22. *Chemical Nomenclature, and Formulation of Compositions, of Synthetic and
    Natural Zeolites; Appendix* No. 41 (1975)

23. *Expression of Results in Quantum Chemistry; Pure Appl. Chem.*, **50**, 75 (1978)
24. *Reporting Experimental Data Dealing with Critical Micellization Concentrations (c.m.c.'s); Appendix* No. 52 (1976)
25. *Selected Definitions, Terminology, and Symbols for Rheological Properties; Appendix* No. 57 (1976)
26. *Report on the Standardization of* pH *and Related Terminology; Pure Appl. Chem.*, **1**, 163 (1960)
27. *Terminology for Scales of Working in Microchemical Analysis; Pure Appl. Chem.*, **1**, 169 (1960)
28. *Recommendations for Terminology to be used with Precision Balances; Pure Appl. Chem.*, **1**, 171 (1960)
29. *Recommended Nomenclature for Titrimetric Analysis; Pure Appl. Chem.*, **18**, 427 (1969)
30. *Recommendations for the Presentation of Results of Chemical Analysis; Pure Appl. Chem.*, **18**, 437 (1969)
31. *Recommended Symbols for Solution Equilibria; Pure Appl. Chem.*, **18**, 457 (1969)
32. *Recommended Nomenclature for Liquid–Liquid Distribution; Pure Appl. Chem.*, **21**, 109 (1970)
33. *Recommended Nomenclature for Automatic Analysis; Pure Appl. Chem.*, **21**, 527 (1970)
34. *Recommendations on Ion Exchange Nomenclature; Pure Appl. Chem.*, **29**, 617 (1972)
35. *Nomenclature, Symbols, Units and Their Usage in Spectrochemical Analysis – I. General Atomic Emission Spectroscopy; Pure Appl. Chem.*, **30**, 651 (1972); *– II. Data Interpretation (Rules 1975); Pure Appl. Chem.*, **45**, 99 (1976); *– III. Analytical Flame Spectroscopy and Associated Non-flame Procedures (Rules 1975); Pure Appl. Chem.*, **45**, 105 (1976); *– IV. X-Ray Emission Spectroscopy; Appendix* No. 54 (1976)
36. *Recommendations for Nomenclature of Thermal Analysis; Pure Appl. Chem.*, **37**, 439 (1974)
37. *Recommendations on Nomenclature for Chromatography; Pure Appl. Chem.*, **37**, 445 (1974)
38. *Recommendations on Nomenclature for Contamination Phenomena in Precipitation from Aqueous Solutions (Rules 1973); Pure Appl. Chem.*, **37**, 463 (1974)
39. *Recommendations for Nomenclature of Mass Spectrometry (Rules 1973); Pure Appl. Chem.*, **37**, 469 (1974); cf. *Appendix* No. 51 (1976)
40. *Classification and Nomenclature of Electroanalytical Techniques (Rules 1975); Pure Appl. Chem.*, **45**, 81 (1976)
41. *Recommendations for Sign Conventions and Plotting of Electrochemical Data (Rules 1975); Pure Appl. Chem.*, **45**, 131 (1976)
42. *Recommendations for Nomenclature of Ion-Selective Electrodes; Pure Appl. Chem.*, **48**, 127 (1976)
43. *Proposed Terminology and Symbols for the Transfer of Solutes from One Solvent to Another; Appendix* No. 34 (1974)
44. *Recommendations on Usage of Terms 'Equivalent' and 'Normal'; Appendix* No. 36 (1974)
45. *Recommendations for Publication of Papers on Molecular Absorption Spectrophotometry in Solution between* 200 *and* 800 nm; *Pure Appl. Chem.*, **50**, 237 (1978)
46. *Recommendation on Nomenclature of Scales of Working in Analysis; Appendix* No. 18 (1972)

47. *Recommendations on Nomenclature for Nuclear Chemistry; Appendix* No. 25 (1972)
48. *List of Trivial Names and Synonyms (for Substances used in Analytical Chemistry); Appendix* No. 45 (1975)
49. *Status of Faraday Constant as an Analytical Standard; Pure Appl. Chem.*, **45**, 125 (1976)
50. *Recommendations for Publication of Papers on Precipitation Methods of Gravimetric Analysis; Appendix* No. 69 (1977)
51. IRVING, H.M.N.H., FREISER, H. and WEST, T.S., *IUPAC Compendium of Analytical Nomenclature,* 'The Orange Book', Pergamon Press, Oxford (1978)
52. *Quantities and Units in Clinical Chemistry (Recommendations 1973) (prepared jointly by IUPAC and IFCC); Pure Appl. Chem.*, **37**, 517 (1974)
53. *List of Quantities in Clinical Chemistry (Recommendations 1973) (prepared jointly by IUPAC and IFCC); Pure Appl. Chem.*, **37**, 547 (1974)

PUBLICATIONS from IUPAC
International Union of Pure and Applied Chemistry

NOMENCLATURE OF ORGANIC CHEMISTRY
(The Blue Book)
Sections A,B,C,D,E,F & H. 1979 revised edition

Editors: **J RIGAUDY**, *Universite Pierre et Marie Curie, Paris, France,*
and **S.P. KLESNEY**, *Dow Chemical Company, Midland, Michigan, USA*

Contents:
Introduction. Hydrocarbons. Heterocyclic systems. Nomenclature systems. Characteristic groups containing carbon, hydrogen, nitrogen, halogen, sulphur, selenium and tellurium. Coordination compounds. Organometallic compounds. Chains and rings with regular patterns of heteroatoms. Organic compounds containing phosphorus, arsenic, antimony or bismuth. Organosilicon compounds. Organoboron compounds. Stereochemistry. Natural products and related compounds. Iṣotopically modified compounds. Lists of names for radicals. Detailed index.

ISBN 0 08 022369 9   550 pages   250 x 176mm

COMPENDIUM OF ANALYTICAL NOMENCLATURE
(The Orange Book) 1978 edition

Editors: **H.M.N.H. IRVING**, *University of Leeds, UK,* **T.S. WEST**, *Macaulay Institute for Soil Research, Aberdeen, UK,* and
H. FREISER, *University of Arizona, USA*

"It should be a part of every analytical chemist's stock of "working books" and placed in a prominent place on the work bench. With over 1,500 definitions, well indexed and readily understood, it is the ideal book for ensuring that ideas are precisely transmitted and that some order comes out of the chaos caused by the haphazard and indiscriminate use of various terms to describe the same thing.
This book cannot have too high a recommendation."   **The Analyst**

ISBN 0 08 022008 0   240 pages   273 x 188mm
ISBN 0 08 022347 8 f

HOW TO NAME AN INORGANIC SUBSTANCE
Guide to the use of NOMENCLATURE OF INORGANIC CHEMISTRY
(The Red Book)

Editor: **W. CONARD FERNELIUS**, *Kent State University, Ohio, USA*

This Guide will aid all who have occasion to use The Red Book to locate more quickly those portions of the rules (indicated by rule number) that are pertinent to their areas of interest.

The combined set of **NOMENCLATURE OF INORGANIC CHEMISTRY** (The Red Book) with **HOW TO NAME AN INORGANIC SUBSTANCE** is available.
ISBN 0 08 021999 3 f

 PERGAMON PRESS

UK   Headington Hill Hall, Oxford OX3 0BW
USA   Fairview Park, Elmsford, New York 10523

Appendix

Important Recent Changes from IUPAC Nomenclature by 'Chemical Abstracts Indexes'

The object of this Appendix is to enable a reader to find easily whether or not the changes in recent *CA* indexes include a specific matter of immediate interest to him. The changes are of two kinds: choice between IUPAC permitted alternatives and adoption of practices not specified in IUPAC rules. Entries in this Appendix all refer to the nomenclature currently used in *CA* indexes, and not necessarily to usages permitted in other CAS publications or to the corresponding IUPAC rules. Fuller descriptions can be found on the pages cited.

'Acid' salts
    Hydrogen is not attached directly to the name of the anion, e.g. sodium hydrogen carbonate, not sodium hydrogencarbonate, 25.

Acyl radicals
    Formyl, acetyl, benzoyl and carbonyl are retained; all other acyl radicals are designated as, e.g. 1-oxoalkyl, 1,n-dioxoalkyl, X-carbonyl (where X is a cyclic radical such as fluorenyl), 1,n-dioxo-1,n-alkanediyl, etc, 110–111.

Alcohols
    Substitutive and not radicofunctional names are used, e.g. methanol and not methyl alcohol, 105–106.

Aldehydes

These are named by use of al or carboxaldehyde suffixes. Only formaldehyde, acetaldehyde and benzaldehyde are used with the simple aldehyde termination, 115.

Alkanes

All are named systematically; the prefixes iso, *neo, sec* and *tert* are not used, 75, 100–101.

Alkaloids

Wide use is made of stereoparents, but very simple compounds are named systematically, 163.

Alkyl radicals

All are named systematically; the prefixes iso, *neo, sec* and *tert* are not used for alkyl radicals, 100–101.

Allyl

Replaced by 2-propenyl, 100.

Amidine

Changed to imidamide, 61.

Amidino

Changed to aminoiminocarbonyl, 61.

Amines

Names ending in amine are formed systematically, e.g., ethanamine, 121.

Amino acids

Thirty stereoparent names are used. Derivatives are named systematically, 161.

Aminocarbonyl

Replaces carbamoyl, 61.

Aminoiminocarbonyl

Replaces amidino, 61.

Ammonium salts, quaternary

Named as derivatives of the 'senior' primary amine, 122.

Benzene derivatives
Trivial names for almost all simple substituted benzenes are abandoned; benzene itself is retained, 76.

Bicyclo [X.Y.Z.] alkanes (X ⩾ Y > Z)
Description of stereoisomerism, 137–138, 141.

Bis, tris, etc.
Used to avoid ambiguity, 46–47.

CA index system
Generalities, 1–3, 42.

Carbaldehyde
Changed to carboxaldehyde, 61.

Carbamoyl
Changed to aminocarbonyl, 61.

Carbohydrates
Trivial names are used only for C_5–C_6 monosaccharides, 152.
Common disaccharides are given systematic names, 155.

Carbonic dichloride
Replaces carbonyl chloride, 105.

Carbonyl chloride
Replaced by carbonic dichloride, 105.

Carboxaldehyde
Replaces carbaldehyde, 61.

Carboxamidine
Changed to carboximidamide, 61.

Carboxylic acids
Only formic, acetic, benzoic and carbonic acid names, and trivial ones for common amino acids, are used by CA; all others are systematic, 109–110.

Cation ium names
These are adopted when the cation is considered to be formed by addition of a proton to an unsaturated compound or by loss of an electron from a free radical, 104.

Chain length
Preferred over unsaturation to decide the main alkyl chain, 58, 69, 71, 101.

Charge on an ion
See Ewens–Bassett system.

Conjunctive nomenclature
Minor variations, 68–69.

Cyclic ketones
See Ketones, cyclic.

Cyclic radicals
Always derived without abbreviation and including indicated hydrogen (if present), 101–102.

Cyclitols and analogues
The method of designating the stereochemistry of substituted cycloalkanes varies with the number of positions substituted, 147–148, 155.

Elements
British–American differences, 7, 9.

Epithio
Treated as non-detachable, 63.

Epoxy
Treated as non-detachable, 63.

Ethenyl
Replaces vinyl, 100.

Ethers
All are named substitutively by use of R-oxy or Ar-oxy prefixes (radicofunctional names are all excluded), e.g. ethoxyethane, 107.

Ewens–Bassett nomenclature
Adopted to denote charge on an ion, 16; examples, 27–29.

Halocarbonyl
Replaces haloformyl, 61.

neo
No longer used as a prefix in aliphatic chemistry, e.g. neopentane is now termed 2,2-dimethylpropane, 75 (cf. Alkanes).

Nitriles and related groups
Nitrile is retained as a suffix, but related groups (e.g. $-OCN$) are named only as prefixes, 114–115.

Nitrogenous cations
Use of ium endings, 23, 121–122.

Non-abbreviation
Names of cyclic compounds and their radicals are no longer abbreviated, 101–102, 117–119

Nucleic acids
The common bases are named as derivatives of purine or pyrimidine, 162.

Organometallic compounds
closo is not used for structure of certain cage compounds, 173.
Coordination names are widely but not universally used, 168.
For cyclic organometallic compounds, Hantzsch–Widman names are preferred to replacement names, 172.
For substitutive names of organometallic compounds the organic part is chosen as parent if it contains a group that should be cited as suffix, 168.
Organoboron and organosilicon compounds: functions are designated by suffixes, 168, 169.
Sulfa, sela and tella are not used in naming rings or chains, 172.

Parentheses
Used around all compound radical names, 47.
Used in 'symmetrical' names, 73.

Partly hydrogenated carbocycles
These are named systematically and not by trivial names, 82.

Peroxyacids
Now named as peroxoic or carboperoxoic acids, 110.

Phenols
No trivial names are used for phenols except phenol itself (C_6H_5OH), 105.

pi (π) complexes
eta (η) is used with modification, 30.

Polymers
Structure-based names are preferred, 176–177.

Porphyrins
These are named as derivatives of 21H,23H-porphine, 164.

Prefixes, detachable
Current usage, 63.

Prefixes, multiplicative
Used for all complex expressions, 46.
Used in 'symmetrical' names, 47, 73.

Purines and pyrimidines
Systematic and not trivial names used, 162.

Quinones
Named as diones, 119.

Radicals, bivalent
ylene and ylidene designations replaced by diyl, triyl etc., except
methylene which is used for $-CH_2-$ and for $CH_2=$, 101.

Radicals, cyclic
Names are always derived without abbreviation and including
indicated hydrogen (if present), 101–102.

sec
No longer used in aliphatic chemistry, 75, 101.

Side chains on rings
Priority is always given to the cyclic component, no matter what the
side chain, 89.

Stereoisomerism
General *CA* procedure, 128, 137, 138, 140–143.
Absolute chirality, 140–143.
Bicyclo[X.Y.Z.] alkanes when X$\geqslant$Y$>$Z, 138, 141.
cis-trans, CA use of, 137, 141.
R*, S*, use of, 140–141, 143.
α, β, extended use of, 141–143, 147.

Stereoparents
Definitions of classes B and C, 146, 152.

Sulfamic acids and sulfamide
Use retained by *CA*, 24.

Sulfoxides
Named by use of X-sulfinyl prefixes, 124.

Superoxide
Supersedes hyperoxide, 16.

Terpenes
Mono-, sesqui- and di-terpenes are named systematically, but stereo-parents are used for tetracyclic and stereochemically complex members, 160.

tert
No longer used as a prefix in aliphatic chemistry, 75, 101.

Unsaturation
Disfavoured relative to chain length for selection of main alkyl chain, 58, 69, 71, 101.

Vinyl
Replaced by ethenyl, 100.

Vitamins
B_6, D, E and K are named systematically, 165.

Vowels
Elision of, 48.

Ylene radicals
No name ending in ylene to specify an aliphatic or cycloaliphatic radical is accepted, except methylene (for $-CH_2-$ or $CH_2=$) and phenylene, 101.

Index

Thioacids, 19, 20, 25
Thioketone, 122
-thiol, 61, 123
-thiolate, 123
-thione, 122
Thionia-, 122
Thiophene, 91
Thioxo-, 122
trans-, 35, 136, 137, 141
Transition elements, 9
Tri-iodide, 16
Trinaphthylene, 81
Triphenylene, 80
Tritio-, 9, 169
Tritium, 9
Trityl, 106
Trivial names, 42, 54, 55, 107
Tungsten, 7

-un, -une, -ur, 151
Uranyl-, 23
Unsaturation:
  conjugated, 75
  cumulative, 76
  role in choice of main chain, 71

Valency, 16
Vinyl-, 100
Violanthrene, 83
Vitamins, 164
Vowels:
  elision of, 11, 48, 75, 86, 95
  insertion of, 48

Water, 14; *see also* Hydrates

Werner, system for coordination compounds, 6
Wolfram, 7
Wolfromate, 20

Xanthene (9*H*-), 91

-yl (free radicals), 102, 103
  (inorg.), 23, 24, 27
  (org.), acyclic molecules, 100, 101
  (org.), cyclic molecules, 101, 102
-ylene, 101
-ylidene, 101
-yne, 103, 136

Z-form, 137
Zincate, 20
Zincide, 13

α,β (alpha, beta), stereochemical infixes, 138, 142, 145, 158
Δ (capital delta), to denote unsaturation in special cases, 96
η (eta) or hapto, prefix denoting multiligancy in a complex, 30
λ (lambda), for citation of connecting number, 172
μ (mu), prefix in names of isopolyanions, 37
μ (mu), prefix denoting a bridging group in a complex, 32
π (pi), prefix in denoting complexes, 29
ξ (xi), stereochemical infix denoting unknown configuration, 146